T0179947

ARE WE UNIQUE?

Also by James Trefil

A Scientist in the City

1001 Things Everyone Should Know About Science

Science Matters
(with Robert Hazen)

Reading the Mind of God

The Dictionary of Cultural Literacy
(with E. D. Hirsch and Joseph Kett)

Space Time Infinity

Dark Side of the Universe

Meditations at Sunset

Meditations at 10,000 Feet

A Scientist at the Seashore

The Moment of Creation

The Unexpected Vista

Are We Alone?
(with Robert T. Rood)

From Atoms to Quarks

Living in Space

ARE WE UNIQUE?

A Scientist Explores the Unparalleled Intelligence of the Human Mind

James Trefil

John Wiley & Sons, Inc.

New York • Chichester • Weinheim • Brisbane • Singapore • Toronto

Copyright © 1997 by James Trefil
Published by John Wiley & Sons, Inc.

Library of Congress Cataloging-in-Publication Data

Trefil, James.
    Are we unique? : a scientist explores the unparalleled
  intelligence of the human mind / James Trefil.
       p. cm.
    Includes bibliographical references and index.
    ISBN 978-1-62045-547-0
    1. Human information processing. 2. Thought and
  thinking. 3. Intellect. 4. Artificial intelligence. 5. Psychology,
  Comparative.   I. Title.
  BF444.T74   1997
  153—dc20   96-34338

10 9 8 7 6 5 4 3 2 1

To Harold Morowitz and the Monday afternoon gang at the Krasnow Institute, who never let me get away with anything.

Contents

Preface

There was nothing particularly unusual about the setting—it was just one more large, white room in a modern building full of large, white rooms. All around me computers hummed as young men and women stared intently at display screens. I was staring at one myself, one I had been directed to by my host, Ryszard Michalski. A slender man with long blond hair and engaging Continental manners, Michalski is recognized as a world leader in a new field of science that bears the innocuous name of "machine learning." His purpose that day was to introduce me to some of the products of his research, beginning with the simple little computer game I was playing.

Here's how it worked: The computer would display a set of figures on the screen. Each figure could have different body or head shapes—circle, square, triangle, etc.—in any of a number of colors. The figure could have a hat or not, carry a flag or not, and so on, and each of these appendages could also be of a different color. Obviously, there were a large number of possible figures, and the screen would show only twenty or so at a time.

With the figures on the screen, a voice came from the machine asking you to make up a rule about what figures would be "in" and which "out." For example, you might decide that only figures with square heads would be "in," or only figures carrying yellow flags. The machine then asked you to tell it a few of the figures that were in and a few that were out. It would then look at the information you gave it and try to figure out what your rule was.

A word about the voice. It was clearly not human, but in a way that was hard to pin down. It wasn't the metallic voice we've become accustomed to associating with robots in movies, but something at once more alien and more difficult to describe. It had a very strange pronunciation. For example, its version of "rule" came out something like "reeool." Not only the accent made the voice sound strange, however—many people speak English with accents more difficult to understand. There was also something about the voice that was different, as if the designers were determined to force you to acknowledge that you were being addressed by a machine, not by a human being.

After you had entered a few "ins" and "outs," the machine would be silent for a moment, then announce, in that uncanny voice, that it had discovered the "reeool." It would then proceed to tell you how you had made your choices. Sometimes, if it didn't have enough information to make a choice, it would ask you to enter a few more "ins" or "outs."

The game was fun. The machine would often come up with a rule that worked for the data you had given it, but wasn't the rule you'd had in mind. If you pushed it, you could often get it to come up with three or four alternate rules for the same set of data.

What was interesting about the machine's responses was that its programs seemed to imitate human reasoning. It seemed to be able to reproduce at some primitive level the ability to deal with ambiguity, with intuition, with all that ill-defined set of abilities that characterize human thought. Confronted with insufficient information, it made a guess. Only if the guesses didn't work did it ask for more information. Would a human behave any differently?

My first reaction was excitement. I was playing a game, of course, but the applications of this sort of system were obvious. As human beings delve deeper and deeper into the world, we use up the simple problems. Increasingly, the problems we want to solve are complex, and it's often difficult to see through the mass of data to the underlying simplicity that we believe is there. The trees make the forest all but invisible. A machine like this could cut through the impenetrable thicket of experimental results and give us a set of rules that might explain them. Once the rules have been called to our attention, of course, verifying and expanding them is relatively easy. It's discovering them that's hard.

In biochemistry, for example, there are a huge number of molecules that operate inside every cell. Are there simple rules that describe their structure and function? We believe so, although we've been able to discern precious few of them. Would a machine like this be able to help? And what about problems in ecology? There may be thousands of variables that describe a particular habitat. Which are important? Which should scientists consider when asked to make assessment of the impact of a new dam or factory? Often, we just don't know enough about the rules that govern the ecosystem to be able to say.

As the afternoon wore on and my probing of the machine's abilities became more subtle, other questions came to mind. "What if I fed all the information about Rembrandt paintings into this machine?" I asked. "Could it tell me how to produce a new one?"

Michalski smiled. "No," he said. "We really didn't know how to deal with the information in a painting."

I was reassured, but only momentarily. The machine I was using was the size of a suitcase—scarcely bigger than the PC on which I'm writing these words. What if someone gave these guys a CRAY? What if we waited ten years and gave them the best machine available then? What if you put a team of hot programmers on the problem for a decade? Would you then have a machine that would tell us how to make a Rembrandt, or worse yet, program another machine to do it?

Suddenly the room seemed a lot less cheery. It was still just as white, and the young people at their computers were still just as earnest and well-meaning. But what were they doing? Was I present at the creation of something that could someday make humans obsolete? Could these machines ever be "human," whatever that means?

I decided to put the machine to one more test. Acting quickly, at random, I fed it a set of "ins" and "outs" with no rule in mind. The machine sat and whirred for a long time, then announced that it had found the "reeool." "The rule is that the figure has either a square body or a yellow body, a hat or a yellow flag. . . ." The machine went on and on, spinning a rule of such complexity that I was flabbergasted. More important, I knew that no human being could ever have found that rule—that I was listening to the words of an intelligence that was totally alien.

As that eerie voice continued, I felt the layers of rationality and civilization that we all erect sliding away, giving way to the primitive fears that lie underneath. I could almost hear the wings of my ancestral vampires flapping in the last rays of a Carpathian sunset. I suddenly knew as certainly as I've ever known anything in my life that I was in the presence of . . . what? Evil? The word seemed both too strong and too insipid to describe what I was experiencing. Then it came to me. I was in the presence of *sacrilege*. What was being done in this very ordinary white room constituted nothing less than an assault on the human soul.

Strengthened by years of training, my rational mind quickly regained control. This was, after all, the last decade of the twentieth century, not some grade B movie. My host was no Victor Frankenstein, and those earnest young men and women weren't assistants named Igor. Great good would undoubtedly flow from their work—perhaps new cures for cancer, new tools in our constant struggle to feed mankind and conquer disease. I spent a few moments chatting with them, discussing the possibility of using their programs on one of my research problems. We arranged for me to give a seminar to explain the problem to the group, and I left.

Outside, in the vinyl-tiled hallway, I paused. I am not a religious man—it has been years since I've been to church. But I want to tell you, my friends, that whether it was rational or not, before I left that place I made the sign of the cross.

1

Is There Anything Left for Us?

What a piece of work is man!
How noble in reason! . . .
In apprehension, how like a god!

—SHAKESPEARE, HAMLET, ACT 2, SCENE 2

Are human beings in some way different, unique in creation and in the eyes of God? Are we, in other words, special?

This is a very old question, and one that at first glance seems to have an obvious answer. Imagine, for example, that you were an extraterrestrial in a flying saucer approaching the planet Earth for the first time. Your instruments would pick up the usual signals of life—water vapor, oxygen, and so on. You'd settle in expecting to find a living planet with a complex ecosystem and then—whoa!—you'd see something absolutely startling. On this particular planet one species completely dominates the ecosystem. It's found virtually everywhere, and its works are of sufficient magnitude that they affect the rest of the planet's systems. By building large-scale reservoirs and lakes, for example, this species has actually slowed down the planet's rotation! It produces works of science and art beyond the capabilities of any other life form. If you knew anything about evolution and natural selection, you'd have to say "This is amazing! Something has happened here. These animals have found a new way to win at the evolutionary game— something no other species on the planet has developed."

Here are some of the characteristics of human beings that a hypothetical extraterrestrial might remark upon: the human ability to transmit nongenetic information from generation to generation via written and spoken language, the ability to devise massive technological systems that produce effects comparable to those produced by natural systems on the planet, the ability to use culture (rather than genetic selection) as a tool in the battle for survival, the ability to develop and manipulate abstract information, resulting in systems such as the ones we call science and language. Depending on its intellectual bent, the extraterrestrial might assign even more weight to the great moral systems embodied in the world's social and religious codes, to the complex aesthetic systems behind the buildings, paintings, music, and literature that permeate human life. All of these might seem to our hypothetical visitor (and to most of us as well) to make a pretty clear case for human uniqueness.

But appearances can be deceiving. It has lately become fashionable among intellectuals to ignore ways in which humans are different from other living things and concentrate on the ways in which we are similar. This trend, I think, may be partly fueled by a misplaced sense of egalitarianism among academics, but it is also based on a lot of good, interesting, and new data. As we shall see in chapter 3, we are starting to learn a great deal about animal behavior. We are starting to find that abilities that used to be thought of as uniquely human—the use of tools, for example, or of language—can sometimes be seen at some level in other living things. As astronomer Carl Sagan and coauthor Ann Druyan put it in their book *Shadows of Forgotten Ancestors* (Random House, 1992):

> Philosophers and scientists confidently offer up traits said to be uniquely human, and apes casually knock them down— toppling the pretension that humans constitute some sort of biological aristocracy.

So one assault on the notion of human uniqueness comes from studies of nonhuman animals. Some of what you read on this subject tends to be overblown, amounting to a claim that because animals can do some things that were previously thought to be uniquely human, there is no difference between humans and animals. I will argue that there comes a point where differences in degree become marked enough to become differences in kind. There is, for example, a rather profound difference between the toolmaking involved in a chimp using a stick to gather termites and that involved in humans building a jet aircraft or a skyscraper.

The traditional response to the question of the difference between humans and animals, of course, was the assertion that only humans possess a soul. In essence, this has the effect of removing the question of human-animal differences from the realm of scientific inquiry, a step I would be extremely reluctant to take.

It is possible, however, to approach this question without giving up either human uniqueness or scientific inquiry. To introduce an analogy we'll be using throughout the book, marking the precise boundary between humans and the rest of the animal kingdom is something like marking the boundaries of a city by traveling out along different highways and noticing where the City Limit signs are. If we pick enough highways on which to travel, and if we carefully note where the countryside ends on each one, then when we connect the dots we'll have a good approximation to the boundary of the city. In the same way, if we consider certain types of ability (the "highways") and look at animal studies, we will be able to find a point for each at which we can say, "Animals can go this far, and beyond here only humans can perform." In the end, we'll have produced a map of those activities and areas that are uniquely human.

The problem until now has been that people have tried to deal with this issue with too broad a brush. The question of whether animals have language ability, after all, isn't one that can be answered "yes" or "no." Instead, we should be asking about what level of language ability can be achieved by what animal under what circumstances, and use that information to "mark the city limits" in this area. When we are done with this process, we will be able to say precisely what separates humans from the rest of the animal kingdom, without necessarily being able to produce broad (and glib) generalities. And if those differences turn out to involve matters of degree rather than matters of kind, so be it. That's the nature of the world in which we live.

Actually, although the issue of animal intelligence is of interest to scientists and philosophers, I don't think most people are very concerned about the fact that some animals seem to have some limited ability to perform functions that most of us think of as uniquely human. Aside from some vociferous rear-guard action on the part of Creationists, most people (religious leaders included) made peace with the fact that human beings are part of the natural world not too long after Charles Darwin published *The Origin of Species*. We recognize that we are part of the great web of life that exists on this planet, and that means that we are related, both by blood and by descent, to every other part of the web. The reason this fact doesn't bother us is that we, like the hypothetical extraterrestrial, can see at a glance that no matter how close that relationship is, there is something *different* about us. And if intellectuals can't define that difference in precise language, who cares? To paraphrase former Supreme Court Justice Potter Stewart when he was pressed to give a definition of pornography, "We know it when we see it."

In fact, we know that this difference has mostly to do with the functioning of one human organ—the cerebral cortex in our brains. In chapter 2 we will explore the connection between *Homo sapiens* to the rest of the web of life and argue that from a biological point of view, it is this organ that provides the difference we seek—that pushes the "city limits" far away from us. Everything else about us, from our skeletons to the innermost working of our cells, is similar (and sometimes identical) to the ordinary run of things in the animal kingdom. As far as the human-animal boundary is concerned, we can rest assured that we are the same, yet different.

I should point out that the notion of human uniqueness is perfectly consistent with modern evolutionary biology. As we shall see in chapter 7, there are many species that have evolved unique adaptations over the millennia—think of the Venus's-flytrap and the bat's sonar navigational system, for example. Being unique doesn't necessarily make you special!

But as you probably guessed from the incident I recounted in the preface, my main concern does not lie with any imagined encroachment of the mental abilities of animals into the human sphere. With all respect to my colleagues in animal research, I don't see the day coming when a chimpanzee will be able to do a calculus problem or compose a symphony, no matter what kind of training it gets. Instead, I'm worried about a very new kind of incursion on traditional human space, one that comes from the machines that human beings, using their cerebral cortices, have built.

The reigning image we have of the human brain these days involves the machine we call the computer. In chapter 9 we will discuss a school of thought called "Strong Artificial Intelligence" (or Strong AI for short). The basic tenet of this school is that the brain is basically the same as a digital computer, although it's obviously more complicated than any computer we've built up to this point. If this is true, the argument goes, then it's just a matter of time before we are able to build a computer that's as complex and sophisticated as the human brain—just a matter of time before everything that our brains can do will be done by a machine. Although I'll argue later that this conclusion is far from obvious, it certainly defines another challenge to human uniqueness.

Go back to the city limit analogy. At any particular time, at any particular level of technology, we can define the boundary between human beings and machines by looking for the point beyond which machines cannot go. The idea would be to define a particular task ("make a painting," for example, or "solve this equation") and see how far a machine can go. On one side of the boundary, the machines can perform as well as (or better than) humans; on the other side, humans still hold sway, at least for the time being.

As we did when we were talking about the differences between humans and other animals, we can use this procedure to delineate a boundary between the domain of humans and the domain of computers. For the sake of argument, let's say the animal-human boundary marks the southern edge of our "city," the computer-human boundary the northern edge.

If the past few decades have seen a slow erosion of the notion that a wide gulf separates us from animals, they have seen a virtual disappearance of the notion that there is any gulf at all between the human brain and computers. You can see this in the widespread (and largely unexamined) assumption that the brain is just a complex computer. This idea takes its most extreme form in the notion that *Homo sapiens* is just a transitory form between the carbon-based life of the past and the silicon-based life of the future. Sometimes this hope leads to wild hyperbole, as when one artificial intelligence enthusiast a few years ago defined the goal of humanity to be "to build machines that will be proud of us."

If the computer jocks are right, if the brain is just a computer that we'll learn to duplicate and improve as time goes on, then the human-machine boundary can be expected to change quickly in the decades ahead. And this prospect, in turn, leads to an important and disturbing question: *When all the boundaries are drawn, when we have understood the limits both of other animals and machines, will there be anything left that is uniquely human?*

In terms of our city limits analogy, when we get through defining the southern border by looking at animals and the northern border by looking at computers, is there any city left in between?

What makes this question complicated is that we are just starting to explore both boundaries. In fact, the explorations are being done by two groups of scientists who almost never talk to each other and often remain blissfully ignorant of each other's work. Zoologists and psychologists make up the main cohort working on the animal side, while computer scientists and systems engineers are looking at the machines. By training and temperament, scientists in these two areas don't mix very well. Biologists and animal psychologists tend to have an abiding appreciation of the complexity and interdependence of natural systems. They are reluctant to make generalizations and tend to get locked into compartments—the insect people don't talk to the octopus people, and both are jealous of the attention vertebrate people get in the press, for example.

On the other hand, the kind of computer scientist who gets involved in these sorts of questions, with a few notable and important exceptions, tend to be "idea people." They can spin off a general theory at the drop of a data point and generalize to all living systems from the results of a single simple computer program. To the computer people, the biologists, with their compulsive attention to detail, are hopelessly stodgy, while the biologists label the computer people with the most withering epithet in their vocabulary—"unsound." As a physicist who has put in a lot of time in the biological trenches, I can appreciate both points of view. Each has a role to play in working out the questions we'll face in this book, each has something important to tell us about our species. If we are to answer the question we've posed about human uniqueness, we'll have to find out what both groups are saying.

From my point of view, the most difficult aspect of the problem of human uniqueness has to do with the possible ability of computers to perform the various functions that we usually put together under the label of "creative" or "abstract intellectual" activities. Could a computer paint the equivalent of the *Mona Lisa*, write the equivalent of *Hamlet*, or produce the equivalent of quantum mechanics or the theory of relativity?

All of these proud achievements are products of the human brain (more specifically, of the human cerebral cortex), so the answer you give to this question depends on two things: (1) what you think a computer can do, and (2) what you think the brain is. After all, if the question is going to revolve around whether the computer is in some way equivalent to the brain, we had better have a pretty good notion about how each one works.

And that brings us to another, more traditional, way of asking the main question of the book. In chapter 5 we will describe at length the main working element of the brain, a type of cell called a *neuron*. The neuron is a physical structure, made up of specific atoms and molecules arranged in a specific way. At present, we don't really understand how the neuron works, but there's no reason to believe that we'll need anything more than the ordinary laws of chemistry and physics to provide such an explanation eventually. A single neuron doesn't "think" and isn't "conscious," at least in the sense in which those words are ordinarily used. Yet the brain, which we believe is nothing more than a collection of neurons, does (again in some way we don't yet understand) produce both thought and consciousness.

In traditional philosophy, a sharp distinction was drawn between "brain"—the physical structure that sits inside the skull, and "mind"—whatever it is that produces the thoughts and mental processes that make up our awareness. As we will stress in chapter 12, a central aspect of human existence is that each of us is aware of an "I" that watches the world go by from a vantage point inside our heads. What is the connection between the physical structure we call the brain, and the mind that we refer to when we say "I"? One way of asking this question is this: *Is the brain just a collection of interacting neurons?*

There are two general classes of answers that people have traditionally given to this question, corresponding roughly to "yes" or "no": I call them mysterian and materialist. As was the case for the human-animal boundary, the extreme forms of these types of answers lead us to conclude either that the essence of humanity must be outside the realm of science or that there is no essential difference between humans and machines.

1. Mysterians

The general tenor of these sorts of responses is that there is something in the human makeup that must forever and irretrievably remain outside of the reach of science, something that can't be explained by the scientific method. I will argue, though, that if you want to claim that there is some sort of "something else" in the brain, then you have an obligation to say what that "something else" is. As I pointed out earlier, the answer to this demand has traditionally been couched in religious terms—the human being, unlike the machine, possesses a soul.

In our secular age, however, this sort of answer just won't do. Although individuals may believe in the existence of the soul, I'm not aware of any serious effort to prove its existence to skeptics. With apologies to my friends who accept the existence of the soul as a matter of faith, I'm afraid it has to be counted out in the general melee of ideas.

Another way to assert an essential mystery to human existence—one that does not depend on religion—is to say that there are some human activities—such as falling in love, appreciating the beauty of a sunset, or helping another with no thought of gain—that will be forever outside the range of machine capability.

I don't have a strong objection to this statement—indeed, I think it is probably true. Proponents of what I will be calling the mysterian position, however, claim that in addition to things like this being unduplicatable by machine, they are essentially different from everything else in the universe—so different, in fact, that they cannot be studied by the scientific method.

Finally, there is an approach, a favorite of New Age types these days, that talks about the mind as a manifestation of some sort of "cosmic consciousness," which, so far as I can tell, they picture as a kind of spiritual miasma pervading our own dimensions as well as all the others they believe exist. I have to say that as an old-line physics prof—a veteran of years of battles against fuzzy undergraduate thinking—this sort of talk just drives me up the wall. As it is usually advanced, with no attention at all to how anyone could prove the existence of this "cosmic consciousness," this notion strikes me as being fuzzy thinking at its worst. I suppose my intense negative reaction to this point of view, at least in part, is fueled by a fear of seeing this book quoted approvingly in some fuzzy New Age publication.

But in the end, I have trouble with the mysterian point of view because it implies that there is a subject of vital interest to human beings—the nature of our own consciousness and mental processes—that must lie forever beyond our reach. *As a scientist*, I can't accept this argument. We've heard this song before. At various times in human history, thunderstorms, volcanoes, disease, the origin of life on our planet, and all sorts of other phenomena were considered to lie in the realm of the gods, outside of human comprehension. But as we got better and better at understanding the physical universe, each of these was seen to be amenable to scientific inquiry. Some, like the question of the origin of life, are far from solved, but the view that the question itself cannot be answered is no longer part of that debate. It's still early in the game for the scientific study of consciousness—way too early to throw in the towel, I think.

2. Materialists

As befits a topic that has been the subject of philosophical debate for millennia, there are all sorts of subtly graded points of view on the connection between mind and brain. We'll encounter some of these in subsequent chapters, but for the moment let me talk about one common point of view that can serve as a proxy for all of the complex and subtle views that have been developed by materialists.

The argument goes like this: The neuron is simply a physical system, composed of molecules and obeying the same laws as other physical systems. Therefore someday we will be able to understand and replicate the neuron. The brain, in turn, is a collection of neurons hooked together. If we can build one neuron, there's no reason we can't build a lot of them, and if we can do that, there's no reason we can't hook them together in complex ways. Therefore, the argument goes, we will eventually be able to build a machine that is a replica of the brain itself. Such a machine would have to have all of the properties of the brain—self-awareness, consciousness, emotions, and so forth. It would, therefore, be "human" in the intellectual sense and capable of doing anything a human being could do. And, of course, if we build this machine, then all we have to do is add more neurons and connections to produce a machine that is, in every sense of the word, superhuman.

This particular materialist point of view, then, starts with the notion that the brain is a collection of neurons, that there really isn't anything else, and goes from there to the notion that someday a machine will be built that will be capable of human thought, human emotions, and human accomplishments.

I guess this is where I get off the train. It's not that I think this chain of reasoning is obviously wrong, although I'll present some arguments later to show that it's not as tight as some people seem to think. The problem is that *as a human being* I am deeply troubled by the notion that all of the great achievements of our species, all the art, all the music, all the literature, all the great scientific insights, are nothing more than the results of the random firing of components of machines that we all carry around inside our skulls. And if I'm troubled by that prospect, you can imagine how I feel about the prospect that all of this will someday be seen as a transitory stage on the road to the supermachine!

Having said this, I have to point out that different people have very different reactions to the notion that someday machines will be able to do everything a human being can. This fact was brought home to me quite forcefully when I was talking about some of the ideas in this book with my two college-age daughters. Confronted with the idea of a machine that could write *Hamlet* or compose the Fifth Symphony, one exclaimed, "But that's terrible," while the other simply shrugged. So if you don't find an intense revulsion welling up inside you at the prospect of a machine producing all the great achievements of the human mind, you're probably in good company. You should then regard what follows as an intellectual exercise concerned with the nature of things such as consciousness, self-awareness, and thought.

Now What?

So here we are. We find the notion of human uniqueness under attack on two fronts—one arising from our membership in the animal kingdom, the other arising from the increased complexity and ability of the machines we build. On each of these fronts, we are confronted with a choice between unpleasant options. Looking at the human-animal boundary, some people argue that we either have to give up trying to make a distinction or have to give up scientific inquiry and accept the existence of something like the soul. At the human-machine boundary, we face a similar dilemma: either accept that the brain is just a reproducible collection of neurons or posit some nonmaterial (and noninvestigable)

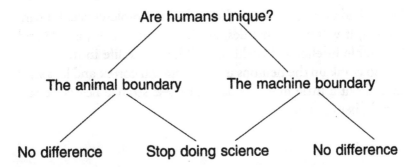

Whether you approach the question of human uniqueness from animals or machines, you seem to have two unacceptable choices: either give up the idea that there is a difference or accept that the difference can't be investigated by the methods of science.

entity. In both cases, the message seems to be the same. Either give up any notion of human uniqueness or give up doing science.

Faced with this choice, I daresay anyone who picks up a book like this will unhesitatingly pick the former option. This seems to be a classic case of Scylla and Charybdis (or its modern equivalent, a rock and a hard place), with the only redeeming feature being that some people don't seem to mind the hard place too much.

But need we give up the notion of human uniqueness so easily? I'm not so sure. In the rest of the book I'm going to tell you why, despite the changes in computer technology and our knowledge of animals, I think there is still room for a human race that is something more than other animals, something unduplicatable by computers.

Make no mistake, though. This isn't going to be a cool, dispassionate examination of an intellectual problem. I desperately *want* to find a way out of this dilemma, and I intend to devote whatever scientific skills I've developed in my career to finding it. If I can't, or if you find my results unconvincing, you'll have another argument about why there's no way out and why human beings had better just resign themselves to obsolescence. If I can, though, it will mean that despite the recent attacks, an old and honorable intellectual tradition still has some life in it.

So crank up the neurons in your cerebral cortex and let's start to find out a little more about what these strange beasts we call human beings are.

2

Humans and Animals
The Same but Different

Let us hope that Mr. Darwin is wrong [about humans being related to apes]. But if he is right, let us hope it does not become generally known.

—VICTORIAN LADY

I think it's possible to make too big a deal about people's dismay at learning that human beings aren't all that different from other animals. It may have come as a big shock to the Victorians when Darwin told them that they were related to the contemporary apes, but my experience with students these days has been that the idea doesn't really bother them all that much. Perhaps it's just a matter of familiarity—modern students, after all, learn about evolution at the same time they learn that the Earth goes around the Sun. In the absence of religious conviction to the contrary, both of these facts seem to grow into features of their intellectual landscape without a lot of fuss.

Nevertheless, if we want to deal with the question of human uniqueness, we're going to have to look carefully at the dividing line between us and the rest of the living things on this planet. A vague "Oh sure, we're part of the animal kingdom" won't do for our purposes. We need to look at that boundary in some detail, and that means that we need to understand where we fit into the great web of life.

There are three ways of approaching this question. We can take the traditional approach, which biologists brought to its fullest flower in the nineteenth century, and compare the anatomy of human beings to the anatomy of other living things. This involves a branch of biology known as *taxonomy*—the science of naming things. Alternatively, we can look at the family trees of living things and ask about similarities of descent, common ancestors, and family relationships. This involves *evolutionary theory*. Finally, we can look at modern humans from a molecular point of view and try to see how our internal chemical machinery differs from that of other living things. This will get us into the forefront of the modern biological sciences—the fields of *molecular biology* and *molecular genetics*.

The verdict of investigations in all of these areas can be summed up in a single statement: Human beings are the same as other living things in many ways, but different in a few crucial aspects. It is in those differences, therefore, that we will have to look for human uniqueness.

"God Creates, Linnaeus Classifies"

If you look around at the living things that you come across on a regular basis, you readily see that it is possible to make a rough ordering or grouping of them. An oak tree is more like a maple tree than it is like a bird, a mosquito is more like a honeybee than it is like a snake, a squirrel is more like a human being than it is like an earthworm, and so on. These kinds of distinctions are obvious, but sometimes they are not so readily apparent—think of the distinction between a whale and a fish, for example. The exercise of making these distinctions greatly preoccupied biologists well into the early part of this century. Faced with the enormous variety among living things, they took it as their first task to try to impose some order on what they saw. One way to do this was to group organisms that had similar functions or structures together. Oak and maples trees, for example, both have a trunk-branch-leaf structure, both have root systems, and both get their energy from processes involving photosynthesis. A bird, on the other hand, has a skeleton and gets its energy by ingesting other organisms. It is reasonable, then, to suppose that in any classification scheme trees would be in one group, birds in another.

The person who gave us the framework of our modern classification scheme was the great Swedish scientist Carl Linnaeus (1707–1778). A physician by training, he became convinced at an early age that his calling was to classify everything on Earth, be it mineral, plant, or animal. (I suppose that you could trace the standard "animal, vegetable, or mineral" scheme of the game Twenty Questions back to him.) As a member of the faculty, he was in charge of the botanical research gardens at the University of Uppsala. Travelers sent him seeds and cuttings from all over the world, and he eventually produced the first general classification scheme for plants—a system that had an enormous influence on the scientists who followed him.

Linnaeus was a strange man. He appears to have suffered from a rather inflated view of his own importance in the grand scheme of things—the quote that opens this section is taken from his work, for example. He made some glaring mistakes (he apparently thought the rhinoceros was a rodent), but he also produced some profound insights. He recognized, for example, that the whale was a mammal and not a fish, and, most important for us, he recognized the close relationship between humans and the great apes.

His most lasting contribution was the use of two Latin names to specify a particular organism. Next time you're at the zoo, take a close look at the signs in front of the cages. There you'll see each animal's common name, followed by two names in Latin— *Ursus horribilis* for the grizzly bear, for example. The first of these names refers to the *genus*, or grouping of closely related organisms in which the animal is found, the second to the *species*, which delineates this particular type of organism. Thus, the genus *Ursus* includes all bears, the species *horribilis* only the grizzly bear.

The system that Linnaeus invented has been greatly elaborated by generations of biologists, but the general strategy is the same. Organisms with similar characteristics are grouped together, and then finer and finer distinctions are made until everything is broken down into distinct breeding populations, or species. As scientists get to the end of this process, there can be a lot of hair splitting in dealing with similar organisms, and seemingly trivial differences—the shape of a crest on a tooth, for example—can take on enormous importance. Going through this process for human beings is one way to see where we fit in the scheme of things.*

The largest clumping of organisms is into kingdoms, and there are generally reckoned to be five of them. *Plants* (which get energy from photosynthesis) and *animals* (which ingest their food) are the most familiar, but modern biologists recognize fungi (which absorb nutrients from the environment) as another, as well as two kingdoms of single-celled organisms (those with and without cell nuclei, respectively).

Among animals, some have spinal cords, and these are in the phylum of *chordates*. Most chordates have backbones, and these are in the subphylum *vertebrates*. Humans are vertebrates. Some vertebrates are warm-blooded, have hair, and suckle their young. These are called *mammals*, and humans are mammals. Some

---

*The expert will recognize that in what follows I have mentioned only a few of the important characteristics of each classification.

mammals have eyes in the front of their heads and grasping fingers and toes. These are the primates. Humans are primates. Among the primates now alive, only one species walks erect and has a large cerebral cortex. This group is *Homo sapiens*—us.

Although this kind of study of classification is no longer at the forefront of the life sciences, there are occasionally surprises. In 1995, for example, scientists found an entire new phylum of organisms that live only on the lips of lobsters!

There is one rather intriguing fact about humanity's place among the animals that we should note here, if only because it helps to explain the long-held human belief that we are, somehow, separate from everything else. If you examine the human family tree, the first thing you notice is that we don't have many close living relatives. Unlike the grizzly bear, who is closely related to all other bears, human beings aren't closely related to anything walking the Earth today. In technical terms, there is no organism now alive that is in the same genus or family as we are— our closest relatives are the great apes, who are rather far from us as these things are measured.*

Things weren't always this way. As little as 35,000 years ago, Neanderthal man was living side by side with modern humans. Whether Neanderthal was a close cousin or a subspecies of *Homo sapiens* is a subject of some debate (I think the data implies that Neanderthal was a close cousin), but he is now gone. At earlier times, several different species of more distant relatives appear to have inhabited the plains of Africa together. Today, however, when we look around us, we see a sizable gap separating us from the rest of creation—a fact that makes it easy for us to imagine that we're not related.

To appreciate this point, think about how the world would appear to an intelligent grizzly bear. Looking around, he or she

---

*Although this situation may explain some of the history of humanity's self-concept, isolation on the evolutionary tree is not at all uncommon among living things. Many species, in fact, have even fewer close relatives than we do.

would see many life forms that looked like grizzlies—polar bears, brown bears, spectacled bears, Kodiak bears, and so forth. It would be much harder for the grizzly to imagine that his species was somehow separated from the rest of living things.

By accident, design, or deliberate action, then, humans have no close living relatives in the tree of life. Nevertheless, when it comes down to stating explicit differences between humans and chimpanzees (our nearest relatives), the list is surprisingly short. The kinds of things that anatomists use to make such distinctions (e.g., the shape of the teeth or the configuration of the sinuses) strike most people as being rather trivial. Even the changes in anatomy associated with upright walking, though very real, don't seem all that important. People have a sense that these sorts of anatomists' benchmarks miss what is *essential* about us.

I suspect that if most people were asked to define the boundary between humans and animals, they would talk about what are generally called higher mental functions—writing novels, producing symphonies, creating scientific theories, and so on. These activities are centered in the brain, and, more specifically, in the outer layer of the brain known as the *cerebral cortex* (or "gray matter"). As we shall see later, most of the kinds of things we point to as being distinctly human arise from the action of cells in the cortex. Thus, from an anatomical point of view, the thing about human beings that matters most as far as distinguishing us from other animals is the existence of a functioning cerebral cortex.

This doesn't mean that other animals don't have cerebral cortices—they do. What distinguishes the human brain is not the existence of the cortex, but its size and organization. If the human cerebral cortex were flattened out, it would be about the size and shape of a large cloth dinner napkin. Our nearest relative, the chimpanzee, has a smaller cortex—something a little bigger than a page of this book. Other animals have still smaller cortices. So when we try to understand the differences between humans and other animals, we're going to have to ask why (and in what way) a fourfold increase in this particular organ can produce such profound changes in behavior. I will argue later that the answer to this question is not to be found in the study of anatomy or even of neurophysiology, but in the new science of complexity.

The Family Tree

The notion that it is the cerebral cortex that defines human uniqueness is borne out by a glance at the evolutionary record—the human family tree. We have only a few fragments of tooth and bone for the earliest hominid. The oldest humans we know most about are members of a genus called *Australopithecus* (southern ape), which first appeared about 4 million years ago. One member of this genus has arguably given us the most famous human fossil ever discovered. I'm talking about Lucy, a member of the species *Australopithecus afarensis* (southern ape from the Afar triangle region of Ethiopia). These early "humans" walked upright, stood about four feet high, lived in social groups, and were most likely covered with fur, like modern chimpanzees. More important from our point of view, they had brains about 400 cubic centimeters in volume—about the same size as adult modern chimpanzees and newborn human infants. Until about 1.5 million years ago, many different species of *Australopithecus* existed simultaneously in Africa.

About 2 million years ago, the first members of the genus *Homo* appeared. *Homo habilis* (man the toolmaker) lasted only about 500,000 years, but *Homo erectus* (man the erect) was far more successful, surviving up until about 500,000 years ago. *Erectus* learned to control fire and spread from Africa around the world. Most of the famous fossils you've heard of—Java man, Peking man, and so on—are from this species. The size of *erectus* brains varied rather widely from one individual to the next. The smallest was about 700 cubic centimeters (almost twice as large as *Austra-*

lopithecus), the largest about 1,200 cubic centimeters (which is well within the range of brain size for modern humans). For comparison, Neanderthal, who appeared only about 150,000 years ago, had an average brain size of about 1,500 cubic centimeters—slightly larger than the modern human average of 1,400 cubic centimeters. Modern humans, *Homo sapiens*, appear in the fossil record about 200,000 years ago.

So when in this family tree do we get to a point where we can say that our ancestors had become fundamentally different from everything else? My own choice would be with *Homo erectus*, mainly on the grounds that there doesn't seem to be a whole lot (other than upright locomotion) that distinguishes *Australopithecus* from modern chimpanzees. This opinion is bolstered by a comment that paleontologist Richard Leakey made in the book *Origins Reconsidered* (Doubleday, 1992), cowritten with Roger Lewin:

> When I hold a *Homo erectus* cranium . . . , I get the strong feeling of being in the presence of something distinctly human. . . . *Homo erectus* seems to have "arrived," to be at the threshold of something extremely important in our history.

To be perfectly honest, I put a higher value on this sort of gut reaction from someone who has lived his life with the fossils than I do on any fancy classification scheme based on measurements.

So, not surprisingly, the evolutionary evidence points to the same conclusion about human uniqueness that anatomy does—what makes humans different from other animals is the human brain. I say this isn't surprising because, to a large extent, both paleontologists and anatomists are really looking at the same things, the human family tree. Paleontologists are busy tracing the branches of the tree, while anatomists concentrate on trying to tell one leaf from another. Small wonder then that you can start from either field and come to the same conclusion.

Having made the point about the brain, however, I should

add one caveat. In the discussion of human evolution, I used overall brain size (measured in cubic centimeters) as a rough gauge of human mental capability. You have to understand that total brain volume is a very rough gauge indeed. In fact, there's no way of telling from a fossil skull how the neurons in the brain were connected—how the brain was wired. As we shall see in succeeding chapters, that's what's most important when we talk about human abilities. The ability to perform functions such as language, mathematics, or music just doesn't show up in a gross measurement like brain volume.

You Are Your Molecules

The great truth of evolutionary science is that all living things are ultimately descended from a single cell that appeared on Earth about 4 billion years ago. The great truth of molecular biology—a truth that has become clear only in the last few decades—is that we carry the marks of this descent in the innermost fabric of every cell in our bodies. A third way to gauge human uniqueness, therefore, is to look at these traces and see whether we can detect any marked difference between us and the rest of creation.

Life is based on chemistry. When we say something is alive, we mean that deep inside its cells thousands of molecules are fitting together, being taken apart, or are acting as brokers while other molecules do the same. The code from which living things produce all the molecules necessary for their functioning is contained in the double helix of the DNA molecules.

Think of DNA as a ladder whose "steps" are made from one of four possible combinations of molecules called *bases*. Everything that differentiates one human being from another, or that differentiates humans from other species, is contained in the message written on those steps in the DNA ladder.

In passing, I should note that the fact that all living things share the same DNA-based genetic code and use many of the same molecules in the most basic cellular operations is strong evidence that all living things descended from a single common ancestral cell.

In human beings, there are about 3 billion steps, or *base pairs*, to use the technical term. About 5 percent of those base pairs carry the information for constructing the molecules that operate in our cells. These stretches of DNA are called *genes*. We don't know much about the part of DNA that isn't in the genes, although it presumably contains (among other things) information about when to turn genes on and off. One of the great frontiers of science these days concerns a detailed mapping of human DNA. Every day new information about the location of genes associated with specific diseases is becoming available, and the most sensational of these discoveries show up in newspaper headlines. The ambitious Human Genome Project is designed to produce a reading of all 3 billion base pairs. It is not unreasonable to think about reading the DNA code to answer questions about differences between humans and other species.

The amount of information in the DNA in a single human cell is about equivalent to that contained in the words of three sets of the *Encyclopaedia Britannica*—there's a lot of information, but not so much that it can't be handled. In principle, then, we could compare two human beings (or human beings and some other species) by laying down the two DNA molecules next to each other and seeing how much difference there is between the messages written in the base pair code. Specifically, we could ask how often the base pair in one DNA molecule is the same as that in the other, and how often the two differ. And even though this hypothetical exercise is a few years (or decades) from being a reality, we already know enough about DNA to make some educated guesses about how such comparisons will turn out when they are finally done.

If we compare the DNA of two human beings, we would find that roughly 1 in 200 of those base pairs would be different, and that the other 199 would be identical. It would be as if we were making a word-by-word comparison of the text of two books and found that the two differed, on average, by two and a half words per page. That's how much genetic similarity there is among different members of the species *Homo sapiens*.

Do the same kind of comparison between human and chimpanzee DNA and you find differences in about one base pair in fifty. To say this another way, human and chimpanzee DNA differs in two base pairs in a hundred, or in 2 percent of the entries. In terms of the book analogy, humans and chimpanzees differ at the level of about ten words per page.

If you try to go beyond our nearest relatives, you run into systematic problems associated with the organisms' different numbers of genes and different amounts of DNA. Because of these differences, it gets hard to know how to line up the two molecules to make the comparison. You can, however, compare common molecules that are coded for by a particular gene to get some sense of how much the DNA of different species might differ, a process that has been done by many scientists. In our book *The Facts of Life: Science and the Abortion Controversy* (Oxford University Press, 1992), my colleague Harold Morowitz and I summarized the results of this exercise for a particular type of molecule called *cytochrome*-C. This is a common workhorse molecule involved in the chemical reactions by which cells generate their energy. In the following table, we show the overlap between this molecule as it is found in other species compared to the version found in human beings.

Organism	Overlap (%)
Chimpanzee	100
Dog	90
Rattlesnake	86
Tuna	77
Pumpkin	71
Brewer's yeast	58

If we assume that these kinds of results can be generalized from one molecule to the entirety of the DNA (and this is a *big* assumption), then this table tells us that we have to get pretty far from *Homo sapiens* before we see significant differences in the coding. In fact, there appears to be more than a 70 percent overlap between human beings and pumpkins, which are not even animals, telling us that we share the great majority of our cellular chemistry not only with other primates but with all living things.

Actually, this result isn't as astonishing as it seems to be at first. Most of the genes in our DNA are concerned with the basic housekeeping of life—obtaining energy, eliminating waste, and so on. That this should be the same for human beings and pumpkins simply says that pumpkin and human cells operate in pretty much the same way at this basic level, using pretty much the same molecules. Given that we both descended from that same primordial cell, this is probably what you'd expect. After all, getting energy from a molecule of glucose involves many of the same chemical procedures, whether that molecule came from photosynthesis (as in the pumpkin) or from a spaghetti dinner (as in humans).

Nevertheless, no one would have any difficulty in telling the difference between a human being and a dog, or between a human being and a pumpkin. The fact that there is so little difference between the respective DNA molecules simply makes a point that we will encounter over and over again in this book: It doesn't take much of a change in an underlying structure to produce major changes at the visible level. In the case that will occupy most of our attention—elucidating the differences between humans and chimpanzees—examining underlying structures like DNA isn't going to get us very far.

Whether we look at anatomy, evolution, or biochemistry, we come to the same result. There are indeed features in human beings that distinguish us from other organisms, but these features tend to be subtle. It is clear that humans are firmly fixed in the great web of life, and that our similarities to other organisms greatly outnumber our differences from them. We are the same, yet we are clearly different.

The differences we recognize as important involve mental activities, the functioning of the human brain. But if we look only at the structure of the brain, as is the case if we look at the structure of the DNA, the difference between humans and other animals isn't all that great. In fact, the difference seems to be a matter of degree rather than a matter of kind. Our task, therefore, is to try to find a way of defining boundaries on what appears, at first glance, to be something of a continuum.

The only way to do this is to recognize that what's important about brains is not how they are built, but what they can do. If both we and chimpanzees have large cerebral cortices, there's not much point in trying to make distinctions based on minute differences in anatomy. Instead, we should look at the end product of the functioning of brains—behavior. It is to this subject that we now turn.

3

Of Fleeing Anemones and Smart Lobsters

All animals are equal, but some are more equal than others.

—GEORGE ORWELL, ANIMAL FARM

What Is Intelligence?

We can't read minds. All we have to go on in judging the mental state of another animal, in fact, is that animal's behavior. If a friend smiles when you enter a room, you assume that your friend is experiencing a pleasurable state because that's what would make you smile in similar circumstances. This sort of extrapolation of mental state seems to work reasonably well when only humans are involved (although even here cultural conventions can produce confusion). It becomes more problematical to apply it to other species—who knows how a deer feels when it stops on the highway to look at you?

When we want to compare human and animal mental states, the only tool we have at our disposal is the observation of what animals do. These sorts of observations come in two general classifications: experiments and field observations. Experiments usually take place in a laboratory environment. They have the advantage of allowing scientists to control the circumstances that affect the animal's behavior and the disadvantage that it is often difficult to interpret the results or know if an animal is motivated to perform. Field observations, as the name implies, involve watching the natural behavior of the animal with as little interference as possible. This technique avoids the artificiality of the laboratory setting but often doesn't allow the control necessary to draw hard conclusions.

In this chapter, we are going to talk about behaviors that usually invoke the word *intelligence*. This is our first, but by no means our last, encounter with a phenomenon that has become all too common in the field of consciousness studies—the appearance of words that most people think they understand, but which have widely different meanings for different people. *Intelligence* can be applied to a phenomenon as simple as a bacterium swimming away from a chemical toxin or to something as complex as the design of an electronic communications system. If what we observe is behavior, the question of whether that behavior implies intelligence is one of interpretation and, in the end, semantics. Rather than get bogged down in a verbal morass at this point, I will use the term in its ordinary colloquial sense. As we go through the animal kingdom, I will tell you what a given animal can do in the way of mental activity and leave it to you to define that ability as intelligent or not.

In this type of discussion, we usually focus on the animal's ability to deal with novel situations—situations that it has never encountered before—with an emphasis on how quickly or how well it learns to cope. The classic sorts of learning experiments involve rats running through a maze to get to food or pigeons in a box learning which button to push to get a reward.

It's important to realize, though, that this way of looking at intelligence contains a strong human bias. We happen to be very good at learning to deal with new situations, so it's perhaps not surprising that we assign intelligence to other animals with the same skill. The reason we're good at this particular task, as we will see in chapter 7, has to do with the fact that our ancestors found that being able to figure things out quickly increased their ability to survive and have offspring. The genetic ability to learn in this way eventually came down to us in our DNA.

There are, in principle, other kinds of intelligence that we tend to ignore because we aren't very good at them. For example, humans are not very adept at paying attention to several things at once—think of the last time you were trying to eavesdrop on two separate conversations at a cocktail party. An extraterrestrial whose ancestors had found this particular trait useful might, in fact, conclude that humans were quite stupid because they couldn't listen to four conversations and two bands at the same time.

The point of this observation is that in what follows I will be primarily concerned with animal behavior that seems to encroach on the areas of mental ability in which humans are proficient. Other animals may not look good in this particular measure but may still be very good at coping with their own peculiar environment. The fact that they can't adapt to new environments simply isn't relevant to their lives, so it's a skill they've never had to acquire.

The field of animal intelligence has boomed during the last decade, and both the number and kinds of different species being tested have expanded enormously. Not so long ago, there was significant data available on only a few organisms. Attention was focused on mammals like monkeys, apes, dogs, and rats, with some data on pigeons (presumably generated because they are easy to keep in a laboratory). Now, however, you can encounter serious scientific debate on the intelligence of the octopus, insects, and even sea anemones!

You might wonder at first why we should be concerned about things like sea anemones if our goal is to search for human uniqueness. The reason is this: By looking at humans as part of the wider web of life, we gain a perspective of the entire intellectual landscape of the animal kingdom. We see human intellect as part of a broad terrain, something we couldn't do if we concentrated narrowly on the differences between humans and their nearest relatives.

Our first job, then, will be to take a stroll through the phyla and see what sorts of behavior different animals are capable of. When we finish, we will find three "bottom-line" truths:

1. It doesn't take a complex nervous system to produce complex behavior.

2. Whatever intelligence is, it's not the private preserve of primates or even of mammals.

3. Nevertheless, it is possible to find a specific point in the scale of mental tasks beyond which only humans can go— tasks that only the human brain is capable of performing.

Intelligence Where You Least Expect It

Consider, if you will, the sea anemone. A distant cousin of the jellyfish, it is often featured prominently in underwater photography because its trunklike body and waving tentacles give it the appearance of a plant, even though it is, in fact, a carnivorous animal. The sea anemone is basically a water-filled muscular bag with one opening where food is moved in and wastes are moved out by the tentacles. It has no sense organs (although it has individual sensory cells) and a nervous system that consists of a network of individual nerve cells. There is no brain, no spinal cord, not even the sorts of concentrations of nerves we call *ganglia*. In fact, its nervous system is pretty primitive by any standard.

Yet in spite of this handicap, the lowly sea anemone is capable of exhibiting a surprising variety of complex behaviors. Ian McFarlane of the University of Hull in England is in the middle of a long study of the sea anemone, its nervous system, and its behavior. His conclusion is that various species of sea anemone can (1) swim away from predators, (2) attack members of their own species who encroach on their territory, (3) climb onto the shells of mollusks, (4) burrow into the sea floor, (5) show an alarm response when a neighbor is attacked, and so on. (In fairness, I should point out that no single species exhibits all of these behaviors, probably because no species has enough nerve cells available.)

Now these behaviors may not sound like high intellect, but they are, in fact, quite complex. Think of what's involved in fleeing from a predator, for example. First, you have to detect the presence of the predator and recognize that a threat exists. Then you have to locate it (so that you know which way to go). Finally, you have to give the appropriate commands to your muscular system to make you move in the appropriate direction. In humans, this sort of behavior is tied up with the functioning of the brain—with recognition of the threat and the control of voluntary muscles. Obviously, in the sea anemone it can't involve a brain, for the simple reason that there is no brain to involve.

Moving up the chain of life, we can look at crustaceans such as lobsters. They *do* have brains, albeit fairly simple ones. Even with this meager endowment, however, they exhibit behavior considerably more complex than that of the sea anemone. Lobsters sense the world around them primarily through the detection of molecules carried in the water. In humans, the ability to detect molecules floating in the air is called *smell*. Someone wearing a perfume, for example, is launching billions of molecules into a room each second, and when these molecules dock on specialized receptor molecules in your nose, you smell the perfume. In the same way, when you see a lobster in a tank waving its antennae around, it is sampling the molecules that are being carried in the water.

Lobsters use their sense of smell to detect small changes in molecular concentrations, and hence to locate the source of odors. They also use molecules as a kind of identification. Special glands secrete molecules into the urine. These molecules, when released, perform roughly the same function for lobsters that the sight of a face does for us—it gives each individual a specific signature that others can identify. The lobster possesses no fewer than three different mechanisms for dispersing chemical signatures into the water, as well as the ability to withhold urine and feces in situations where a predator is near. (This last is the olfactory equivalent of a rabbit freezing when it is being stalked by a predator that locates it by sight.)

So what kind of behavior do we find in organisms that "see" the world primarily by smell and have smallish brains? Although they are essentially solitary animals, lobsters exhibit a rather complex set of social behaviors. Male lobsters fight for the right to inhabit the largest caves, for example. During the fight, both lobsters release urine, and the loser signals the end of the fight by stopping his release (a behavior that caused one participant at a conference on animal intelligence to remark that lobsters teach us that "it is better to be pissed off than pissed on"). For weeks after a fight, the loser avoids the winner, a fact indicating that one lobster can recognize another and modify his behavior according to that information.

Lobsters also execute complex search patterns when trying to locate food. Jele Atema and his research group at Boston University found that they could mimic these search patterns with a simple robot. They built the robot with two chemical sensors, one on each side, and a program that told the robot to swim toward the side where the concentration of a particular chemical was stronger. Placed in a tank with a chemical source, the robots swam in circles, cast around, and eventually homed in on the source, just as a lobster would. Perhaps it was the combination of this result with the complex social behavior that caused Atema to comment, "Sometimes I think of (lobsters) as little people in hard shells, sometimes I think of them as little robots." But whether you think of the lobster as a robot or as a conscious being, the point we made above is reinforced: It doesn't take very much in the way of a nervous system to produce highly complex behavior.

If you want to talk about intellectual achievement, though, it's the octopus that is the Einstein of the invertebrate world. The octopus has a well-developed eye and a relatively large brain. With some 500 million nerve cells, it has the largest nervous system of the invertebrates. It is also the only invertebrate that routinely hunts vertebrates like fish, a point that is made with great glee by octopus people at scientific meetings.

In passing, I should mention that I have always been amazed at how scientists who study a particular species develop proprietary attitudes toward it, and how easy it is to get them going. This remark about octopus hunting habits, for example, seemed to be a source of great offense to the bird and insect people at a meeting I attended recently. Don't ask me why.

In any case, observations of the octopus in the wild show behaviors that clearly limit our evaluation of its intelligence. If an octopus sees a crab run into a hole in a rock, for example, it will use its arms to cover all the holes in the area, then search them one by one, as if it couldn't remember which one the crab entered. If a bit of food is moved while an octopus is reaching for it, the octopus cannot do a midcourse correction or simply move its extended arm to the new position; it must bring the arm back, recalculate, and start the whole reaching sequence over again.

In the early 1900s, a whole series of classical learning experiments were done to try to establish how much the octopus can learn. They were very traditional in design—the animal would be presented with two shapes (a square and a triangle, for example) and then be given food if it reached for one and an electric shock if it reached for the other. The folklore about octopus intelligence largely comes from the experimenters' reports that the animals learned to distinguish various geometrical shapes, different orientations of those shapes, and even different textures of the material from which the shapes were made.

More recently, however, scientists have started to reevaluate the original experiments on octopus learning. It has, after all, been over twenty-five years since anyone has done serious testing of this sort, and we've learned a lot about how these tests ought to be done. For example, when the shapes were presented to the octopus, it was generally done in such a way that a piece of fish was tied behind one and an electrode was attached to the other. It was possible, therefore, that what was being tested was the animal's ability to sense those things rather than its ability to distinguish shapes. In the words of Jean Boal of the University of Texas, "Not all of those experiments come up to the standards of modern experiments with mammals."

It is unlikely that this debate will dethrone the octopus from its place at the summit of invertebrate intellects, although it may lead us to conclude the invertebrates aren't as smart as we used to think. With the octopus, however, we are starting to encounter animals with large, complex brains and the beginnings of what is normally called intelligence. In the octopus, in other words, we see both complex behavior and a complex nervous system. It is probably no accident that this should occur in an animal that learns about its environment through the sense of sight and has to move around to catch its food. As we shall see later, visual processing and the control of movement take up a good fraction of the capacity of the brains of higher animals, including humans.

There is an important lesson we can learn from this stroll through the invertebrates. As we've already noted, simple nervous systems can produce complex behavior, and it appears that adding relatively small numbers of nerve cells (as in the transition from sea anemone to lobster, for example) can produce enormous changes in the ability of an organism to learn and cope with new situations. Therefore, whatever types of abilities we want to describe as "mental capacity," we have to realize that they may depend on relatively small changes in the structure of the brain. To put this insight in a different language, it appears that large differences in mental capacity need not be correlated with large differences in things like brain size and number of nerve cells, or even with large differences in brain structure. As we try to establish the boundary between human and animal capabilities, we will want to pay more attention to behavior, which presumably reflects structure in the brain, rather than to the details of the structure itself.

Animals Like Us

The ultimate task for finding the human-animal boundary, of course, is to understand the difference between us and those animals that most resemble us. And this, in turn, means that we have to think about what makes humans different from other primates and, particularly, from chimpanzees, who are our closest relative on the tree of life.

Scholars have advanced three different ways in which humans may be different from chimpanzees:

1. Only humans make tools.
2. Only humans possess language.
3. Only humans can form mental concepts of certain levels of abstraction.

I've already mentioned in chapter 1 that the first of these statements is no longer considered to be strictly true. Chimpanzees in the wild have been observed to take a long stick, strip off its small branches, insert it into a termite nest, and then eat the termites that cling to it when it is pulled out. There is some evidence that they will also use rocks to break open nuts. There are even reports that crows in New Guinea fashion hooks from thorns to pull insects out of crevices. These are, indeed, examples of toolmaking and have caused some commentators to announce gleefully that differences between humans and other animals are "only a matter of degree."

I have to say that I find this argument distinctly under-whelming. In logical terms, the argument goes something like this:

1. A stick is a tool.

2. A 747 (or a supercomputer or the Empire State Building) is a tool.

3. Therefore, the difference between a stick and a 747 is only a matter of degree.

This sort of argument makes a nice debating point but uses a kind of verbal obfuscation to hide a very important fact. In any situation, there is a point at which differences in *degree* become differences in *kind*. A single raindrop, for example, is fundamentally different from a raging flood, even though both are made from water. A philosopher who stood in the path of a flood and declared that it differed from a raindrop "only in degree" would quickly be made aware of this fact. In the same way, I would argue that anyone who calls the difference between the ability to build a 747 (or even the ability to build a fire) and the ability to use a stick "just a matter of degree" is being willfully obtuse. I would keep toolmaking, then, as one trait that distinguishes humans from even our nearest primate neighbors.

The second distinction—language—is sufficiently rich and complex that I will devote the entire next chapter to it. The bottom line, though, is that if we understand the modern view of what human language is, we will see that it is different both in degree and kind from the communications between one animal and another, as well as the communications between animals and human beings.

The final distinction, the ability of humans to form certain kinds of abstract mental concepts, can be made as a result of experiments that have been done in the last decade. A large part of the motivation for these experiments (as well as the inspiration for the experimental design) comes from attempts to learn how small children form their ideas about the world. What makes experiments on animals difficult, of course, is that unlike children, they can't tell you what they think about something. Consequently, a lot of thought has to go into designing experiments in which the animal's state of mind can be inferred from its behavior.

Take a fundamental concept like the one developmental psychologists call self-recognition. This is defined operationally as the ability to recognize oneself in a mirror—an ability that human children acquire at the age of eighteen to twenty-four months and that seems to be related to the onset of self-conscious emotions and behaviors.

Most animals can't seem to grasp the notion that the image in the mirror is connected to them and is not another animal. This fact was brought home to me forcibly one spring when I was living in the Blue Ridge Mountains. A male cardinal had established a territory near our house, and every afternoon, when the sun was right, he would engage in a furious attack on his image in our living room window. Obviously, he saw the image as a competitor for his territory. To save my window, I took a side mirror off an old pickup truck and bolted it on the wall of the house. The bird then spent his time attacking the mirror, leaving my window alone. (To round out the story, this particular bird stayed around only one year and was replaced the next year by a bird who either had poorer eyesight or was less aggressive.)

The way that animals are tested for self-awareness is simple. First, they are exposed to mirrors until they are accustomed to them. Then, while they are asleep, someone sneaks into the cage and paints the top of their heads (or some other part that can't be seen) a bright red. After that, the animal is observed the next time it goes by the mirror. If it stops, does a double take, and starts to rub the red spot on its head, it's safe to conclude that the animal has formed a mental connection between the image in the mirror and itself. If it treats the image as it always has, on the other hand, it's safe to assume that the particular connection hasn't been made.

This sort of experiment has been done on a wide variety of animals (including, believe it or not, Indian elephants!). The results are unambiguous. Of all the primates, only chimpanzees and orangutans are capable of forming mental concepts of self-awareness (as defined by the mirror test). Other animals that we normally regard as intelligent—gorillas and rhesus monkeys, for example—don't seem to have this ability. Thus, this simple experiment allows us to start drawing lines in the animal kingdom based on the ability to perform a specific mental task. Humans, chimpanzees, and orangutans can exhibit self-aware behavior, other animals can't. Period.

We can design an experiment to test another aspect of mental sophistication—the ability to see the world through the eyes of another—by building the experiment around a simple laboratory game. A situation is set up in which the first player has information, and a second player has to learn to follow the signals of the first to get the reward. For example, the first player may be able to see which of several boxes contains food but not reach the levers that allow someone to get at that food. The second player can reach the levers, but can't see into the boxes. Over a period of time, a chimpanzee or a monkey will learn to pull the lever on the box to which the human being (the first player) points.

But what happens if we now play the game with the players reversed? What if, in other words, we set things up so that the chimpanzee or monkey can see into the boxes. Will it learn to signal to the experimenter to get the food? When this experiment is done with chimpanzees, they "get it" quickly, learning to win the game from their new position in a much shorter time than it took them to learn it from the old. They seem, in other words, to be able to see the game from both sides—to understand and assume the role of both players at once. Rhesus macaques, on the other hand, can't do this. If they are put in the other position, they have to learn the game from scratch.

So once again, we can separate humans and chimpanzees from

the rest of the animal kingdom on the basis of their ability to perform a specific mental task. Can we take this notion further and find tests that allow us to draw this sort of distinction between humans and chimpanzees? The answer turns out to be yes.

The area of research that allows us to draw this sort of distinction goes by the name of "Theory-of-Mind" experiments. The purpose of these experiments is to probe the ability of primates (including human children) to understand that other primates have minds like their own. To be precise, these experiments test the proposition that chimpanzees or children can understand that other beings have minds that contain specific information.

Again, the approach is through a laboratory game. As in the two-way game, a situation is set up in which the chimpanzee is supposed to pull a lever to get food in a situation where it can't see into the boxes. This time, however, there are two people on the other side. The first of these leaves the room while food is being put into one of the boxes, all of which are then covered. The first experimenter then returns. When the game starts, each of the two experimenters points to a different box. The question then is simple: Which set of directions will the chimpanzee follow?

Obviously, what is being tested here is whether the chimpanzee can understand that the other players have different mental states, and that only one of them can have the information needed to get the food. In a recent series of experiments, the chimpanzees started out by opening random boxes, but eventually three out of four began to follow the directions of the person who stayed in the room. The process by which they "got it" followed a classical learning curve. The obvious interpretation is that they were learning a new game whose bottom line was "follow the directions of the guy who stayed in the room."

If, on the other hand, you play this same game with a four-year-old human child, the results are strikingly different. The child doesn't have to go through a long learning process or a period of trial and error—it plays the game correctly right from the start.

The child, in other words, seems to be able to look at the situation, understand that only one of the experimenters has the information needed to complete the game, and follow the instructions of that experimenter. The child "gets it" at once. In the jargon of the experimenters, the child forms a "Theory of Mind" that tells him or her how to play the game, while a chimpanzee forms no such theory and learns to play this game as it would any other—by trial and error.

A similar experiment with chimpanzees makes this point more directly. Instead of having one experimenter leave the room, as above, he or she is simply blindfolded. In this case, chimpanzees seem to follow the directions of the blindfolded and unblindfolded experimenters with equal frequency. Again, they don't seem to be able to form the concept that the blindfolded person can't know where the food is. (You might think that the problem here is that the chimpanzees don't know that a blindfolded person can't see. We know, however, that chimpanzees understand the connection between the eyes and vision. Like human infants, they exhibit "gaze following"—if someone stares in a particular direction, they will start to look in that direction as well.)

Singling Out Human Beings

Looking at animal behavior, then, we can discern a set of constricting circles, each including fewer species than the last. We can recognize predators and flee from them, but so can sea anemones. We can recognize individual members of our species, but so can lobsters. We can perform simple learning tasks, but so can the octopus (not to mention the pigeon and laboratory mouse). We can recognize ourselves and see situations through the eyes of another, but so can chimpanzees. It is only when we get to the ability to form the concept of another's mental state—the ability probed in the Theory-of-Mind experiments—that we find a circle that contains only our species. In terms of the highway analogy used in chapter 1, this series of experiments marks out one point on the boundary between humans and nonhumans, and marks it out with a fair degree of precision. Presumably, future experiments will map the rest of the boundary in more detail.

Thus the conclusion of behavioral studies is the same as the conclusion we reached in chapter 2 on the basis of anatomy and biochemistry—whatever it is that separates us from the other animals has to do with the functionings of our brain. In that three-pound mass enclosed in the bones of our skull lies the secret to human uniqueness.

4

Can Animals Talk?

It seems this guy walked into a country store and
saw a dog playing checkers with the storekeeper.
"What a smart dog," he exclaimed.
"Oh, he ain't so smart," the storekeeper replied.
"I can beat him three games out of four."

—TRADITIONAL STORY

Clever Hans

At the turn of the nineteenth century, a rather strange chain of events transpired in Germany. A retired schoolteacher named Wilhelm von Osten set out to teach Hans, his horse, to do arithmetic. So successful was he that he soon found himself on tour, performing for delighted audiences. The routine would go like this: Von Osten would ask Hans something like "How much is two plus three?" Hans would then start to paw the ground with his hoof—one, two, three, four, five times. Then he would stop. What's more, Hans was able to deal with complex questions. "How many umbrellas are there in the room, Hans?" or "What is the

45

date next Thursday?" Invariably, Hans would tap out the answer. What more proof of animal intellect could you want? Here was a horse that was not only smart enough to do arithmetic, but could communicate his answers in a meaningful way to a rapt human audience! Observers at the time compared him to a smart fourth grader, and gave him the name "Clever Hans" (*der kluge Hans*).

Alas, it was not to be. The horse became so famous that in 1904 the German Board of Education set up a commission to study him. They quickly found that there was no obvious chicanery involved—von Osten was clearly an honest man (for the record, he never charged admission for Hans' performances). Nevertheless, some simple tests began to show that all was not as it seemed. Hans was asked to read a number written on a card. If he could see the questioner, his answers were correct over 90 percent of the time. If the questioner stood off to the side while the horse wore blinders, however, his accuracy rate dropped to 6 percent. Close observations of the questioners finally provided the key to the Clever Hans phenomenon. It turned out that when people asked Hans a question, they would bend over slightly to watch his foot. When he got to the right answer, every observer would unconsciously give a little upward jerk of his head. No one was aware of doing this, but clearly Hans had learned to recognize this movement.

Von Osten was as surprised as anyone by this result—there had been, after all, no attempt to hoodwink anybody. Nevertheless, the incident cast a pall over the field of animal-human communication for generations. Even today, researchers in the field have to be very careful that they're not simply reproducing what has come to be called the "Clever Hans effect."

Having said this, however, it seems to me that one very important point is often missed in discussions of Clever Hans: He had to be one very smart horse to learn to read unconscious signals from his trainers. That he couldn't do arithmetic as well shouldn't obscure that simple fact.

Three Ways of Asking the Question

The use of language has often been advanced as one of the things that divides humans from the rest of the animals. The question of how, and in what way, animals communicate therefore becomes an important issue in delineating the human-animal boundary. There are actually three different questions hidden in this simple query. It's important to realize that they are distinct and discrete, if only because they are often confounded. The three questions are listed below, along with a parenthetical statement of the answers I will try to support in the rest of this chapter.

1. Can animals communicate with each other? (*Absolutely.*)
2. Can animals and humans communicate with each other? (*To some extent.*)
3. Can animals learn human language? (*Probably not.*)

It is the last question that is most interesting for us, and also the one that is the most controversial.

Animals Talking to Each Other

The vervet monkey spends most of its life as a member of a social group on the African savannah and in the neighboring forest. It's an environment full of danger for small animals, with more than its share of predators. Like many social animals, the vervet monkey has developed a warning system wherein if one member of the group spies a danger, all the rest are warned. If one monkey spots a snake, leopard, or eagle (the vervet's main predators), it screams to warn the rest of the group.

For a long time it was assumed that this scream was simply a startle response—something akin to the way teenagers scream at a particularly scary part of a horror movie. In the late 1960s, however, a group of researchers from Berkeley observing the monkeys in their natural habitat realized that the "scream of terror" was actually three different cries, and that the monkeys' response to each type of cry was different. When they were on the ground and heard the "snake cry," for example, they would stand up and look around at the ground. The "leopard cry," on the other hand, sent them out to the smallest branches of nearby trees, while the "eagle cry" sent them into bushes or thick foliage.

This was one of the first indications scientists had that monkeys are capable of conveying specific, detailed information (as opposed to general emotional states) to each other. The conclusion is truly inescapable—standing up and looking around is clearly not very useful when there's an eagle in the sky. Other studies of animal communication have produced similar types of results, all based on observing what animals do after some sort of communication has taken place. Here are a few examples:

• Honeybees returning from a source of nectar communicate the location of their find to fellow hive members by doing a little dance. If the source is within 30 feet of the hive, the bee dances around in a circle; if it is farther away the bee does a kind of tail-wagging figure eight. The rapidity with which the bee repeats the circuit indicates how rich the source is, while for distant sources the angle of the plane of the dance indicates direction (as reckoned from the location of the sun).

• Male songbirds sing to announce their availability for mating and to keep competing males from their territory. Most songbirds sing several different songs.

• Dolphins make a number of sounds—whistles, clicks, and grunts. Some of these are used to locate objects in the water (think of it as an animated version of submarine sonar). The whistles, however, seem to be used for the identification of individuals. The animals seem to go through life saying "I'm Suzy," "I'm Suzy" to other members of their group. In effect, dolphin communication seems to be the marine equivalent of those little sticky tags they hand out at conventions—the ones that announce "Hello, my name is . . ." (I should point out that as a result of extensive research scientists no longer accept the notion that dolphins are in some way more intelligent than other animals.)

• Whale songs—particularly those of the humpback whales—are complex communications that can be up to twenty minutes in length. Members of the same group of whales all sing the same song, but the songs evolve over time. No one knows why whales sing, although their songs seem to be involved in their mating behavior.

• Wolves routinely communicate complex things relating to social status, such as submission and dominance, to each other through a series of postures. The postures are so clear that they are even well-known to humans.

Some of these communication techniques are instinctive and require no learning. The honeybee, for example, needs no lessons to execute or understand its dance. This particular language is obviously carried from one generation of bees to another in the genes. In other cases, animal language seems to arise from both genetically transmitted information and environmental learning. One way to test this statement with songbirds is to raise the birds in an environment in which they do not hear the songs characteristic of their species. Some species, such as flycatchers, can produce their songs even when raised in acoustic isolation. Others, such as wrens, must have a model from which to learn. In an experiment with cowbirds, for example, chicks from North Carolina were raised around Texas adults. The result: The chicks grew up singing with a strong Texas accent!

Clearly, there is a genetic component to language ability in animals. We shouldn't be too surprised, then, to find a similar genetic component in human language as well.

Humans Talking to Animals

I have been around dogs all my life, so I know from personal experience that communication between species is possible. Anyone who has been to a canine obedience class (or, better yet, sheepdog trials) knows that dogs can understand, interpret, and act upon commands from humans. In a similar vein, anyone who has been to one of the numerous Sea Worlds that dot the landscape knows that porpoises and seals can do the same. Communications from humans to other species is an everyday occurrence, hardly worth commenting upon.

In the same way, animals can communicate with humans to some degree. To take dogs as an example again, most humans can distinguish readily between an approach by a friendly dog (head up, tail wagging, loud bark) and an unfriendly one (head low, bristles up, low growl). We all recognize what ethologists call a "play bow" (rear up, tail wagging, front legs on ground from elbows forward) and know how to respond to it. We can even, with some experience, learn some simple canine etiquette. When a friendly dog approaches, for example, we practice interspecies courtesy by holding our hand out for the dog to smell before we pet it—a simple accommodation to the fact that the dog's view of the world is much more olfactory than ours.

I should point out, however, that it is all too easy for people to fool themselves into thinking that because we can communicate or establish some sort of relationship with an animal, that the animal must somehow think and see the world as we do. *Nothing could be further from the truth!* Except for a few species such as dogs, with whom we have enjoyed a long association, the minds of other animals are fundamentally foreign to us. You can see evidence for this in the many stories of people who raise wild animals from birth, only to have the animals turn on them one day without, so far as the human can tell, any provocation whatsoever. Even the chimpanzees discussed later, raised with humans from birth and brought to the brink of language acquisition, remained wild animals. This fact was brought home to me forcefully over late-night beers with a group of animal behavior researchers recently. The talk turned to chimpanzees (and particularly to Kanzi, about whom we'll talk in a moment). The group started listing all of their colleagues who had had fingers and miscellaneous parts of their anatomy removed in vicious and (from the human point of view) unprovoked attacks during normal activities. The list of victims was longer than I like to think about.

Animals Talking to Us

If you want to talk about animal acquisition of human language, you have to answer two questions: (1) What exactly is human language? and (2) How much of human language can animals actually comprehend and use? Let's look at these two questions in order.

What Is Human Language?

At first, the question of what human language is may seem strange. We use language so offhandedly, so unconsciously, that it is something of an effort to think about it. But starting in the 1960s, our understanding of human language went through a profound change. This change shook the foundations of academic linguistics but remained largely invisible to the public (and, I might add, to academics outside linguistics). The bottom line of the revolution is this: Human language ability seems to be hard-wired into our brains. It is, in other words, a physical adaptation of our species to the environment in which our ancestors found themselves.

The first reaction of most people to this claim is disbelief. After all, human beings speak thousands of different languages. Anything whose manifestations show such variability from one culture to another must surely be the result of social learning rather than some innate genetically controlled wiring in the brain. But consider, if you will, the following observations:

1. Children the world over begin to acquire language at the same time. Babies begin to babble at seven or eight months, using the same sounds regardless of the language being spoken around them. Deaf children whose parents use sign language even babble with their hands!

2. Children acquire language in a well-defined sequence. English speakers, for example, acquire the *a* sound before *i* and *u*, the *p*, *b*, and *m* sounds before *t*. Around their first birthday, babies begin to acquire complete words. All of this (and much more) seems to happen regardless of the child's environment or the specific language to which the child is being exposed. It doesn't seem to depend on the child's motivation or intelligence, either.

3. Language acquisition is extremely rapid. By age six, most children speak in complete grammatical sentences in their native language. Children who do not acquire language by age six have great difficulty in speaking in later life— the longer the delay, the worse the problem. One outcome of this fact is the well-known difficulty adults have in acquiring a foreign language.

4. According to some estimates, the average American high school graduate knows about 45,000 words. If we assume the graduate is eighteen and started learning words at age one, this works out to about 2,600 words learned per year, 7 words a day, or *a new word every two waking hours for seventeen straight years!* That, my friends, is rapid learning, and trying to imagine how language acquisition could proceed without some sort of genetic grounding is difficult.

Given these facts, the idea that there might be some sort of innate human ability to acquire language seems less farfetched. But the real evidence for the innate nature of language comes from the realization, most often associated with the name of MIT linguist Noam Chomsky, that all human languages share the same set of deep grammatical rules. In fact, MIT scholar Steven Pinker, in his marvelous book *The Language Instinct* (William Morrow, 1994), goes so far as to say that from the Chomskian point of view, a Martian scientist visiting Earth would conclude that "aside from the mutually unintelligible vocabularies, Earthlings speak a single language."

The rules of human language are not about sounds or vocabularies, but about the way languages are put together—the way humans invest a particular sound sequence with meaning. These sorts of rules tend to be stated in the language of linguistics (which, mercifully, most of us have forgotten if we ever learned it), and tend to be of the "if-then" type—if a language has property A, then it will also have property B.

To understand an example of these rules, we need a bit of background. In many languages endings are added to nouns to show how they are used in a sentence. For example, starting with the basic word *car*, we say "cars" to indicate more than one car, and "the car's door" to indicate that the door belongs to (i.e., is possessed by) the car. These are examples of what are called *inflections*. As it happens, English is relatively impoverished in inflection—plurals and possessives are about all there is.

Not all languages work this way, as generations of students have learned to their dismay. German distinguishes among masculine, feminine, and neuter nouns, and has four different endings for each type of noun to indicate how it is used in a sentence. Czech also assigns gender to its nouns, and has seven sets of endings. I have been told that Hungarian (a non–Indo-European language) has twenty-three different sets of endings. In Czech, noun endings designate such things as whether the noun is the subject of the sentence ("The car is red"), the object of the verb ("I push the car"), or the indirect object ("Give the car a checkup"). There are also different endings if the noun indicates location ("The hat is in the car"), signifies instrumentality ("I went there by car"), or even whether the noun is being used in address ("Hello, Car"). In English, we use "car" for all of these and use word position to indicate the function of the noun, but in Czech the noun would have a different ending in each situation (e.g., "The car is red" but "The hat is in the careh"). Similarly, in many languages there is a way of changing a verb into a noun—"jump" becomes "jumper" in English, for example. The "er" here is called a *derivational ending*.

Here is a simple example of a rule about how words are put together: If a language has both inflectional and derivational endings, the derivational ending will come before the inflectional in a single word.

An example of how this rule plays out in English is that we say "jumpers" instead of "jumpser."

Now there is no logical reason why a construction like "jumpser" couldn't appear in some language somewhere. It gets the point across as well as "jumpers." Yet the fact is that *no* human language allows this construction! The linguist would claim that the reason that "jumpser" seems so fundamentally wrong to us is that it violates the innate rules of grammar in our brains. Chomsky delighted in illustrating the point with this made-up sentence: "Colorless green ideas sleep furiously." This sentence makes no sense at all, but "feels right" to an English speaker. This is because the arrangement of words conforms to the deep rules of grammar. On the other hand, the equally meaningless sentence, "Furiously sleep ideas green colorless" is gibberish because it doesn't correspond to those same rules.

There are deep rules for things like the use of noun and verb phrases, the use of prepositions (and "postpositions," which don't exist in English but do in other languages), the way words and phrases move around in sentences, and so on. The idea is that human language consists of two levels—the deep level of hard-wired (i.e., genetically determined) rules and the surface level of spoken and written language. What happens when language is acquired is that the baby fits the vagaries of the language he or she hears being spoken into the framework of grammatical rules that are already wired into his or her brain. This scenario is certainly the simplest explanation for both the common structure of human languages and the sequence of human language acquisition.

For, as we pointed out above, children everywhere go through the same sequence in this acquisition. A child goes from babbling to single words to two-word sentences and then, all of a sudden, fluent grammatical speech. It is this sudden onset of grammatical speech that will concern us most. One way of explaining it is to say that the discontinuity corresponds to the advanced grammar "circuits" coming on line. In the 1970s, psychologist Roger Brown published some landmark studies on language acquisition in children that illustrate this transition. Here are sample sentences from one such child—sentences that will resonate with anyone who has gone through this process with his or her own children.

two years, three months: Play checkers. I got horn.

two years, six months: What that egg doing? I don't want to sit seat.

three years: I going come in fourteen minutes. Those are not strong mens.

three years, two months: What happened to the bridge? Can I put my head in the mailbox so the mailman can know where I are and put me in the mailbox?

Language acquisition, in other words, seems to be something like the onset of puberty. Different children go through it at different times, but when it happens it happens fast. Children start to speak in complex sentences, use phrases embedded within each other, and, in general, sound like adults. *All this happens without specific training.*

The idea that an innate set of grammar rules exists is certainly the simplest hypothesis capable of explaining all these different kinds of regularities in language and language acquisition. In later chapters we'll talk about where, precisely, in the brain these circuits might be and how language ability might have evolved in humans. As far as delineating the human-animal boundary, though, what we really want to know is how far along this track of language acquisition, from babbling to grammatical speech, an animal can proceed. In particular, can animals get past the "Big Bang" that occurs when the grammar circuits kick in?

What Can Animals Do?

At the level of naming things, recognizing words, and being able to answer simple questions, there's no doubt that animals can function in the verbal sphere. Perhaps the most striking example of this rule is an African gray parrot named Alex. The protégé of Irene Pepperberg of the University of Arizona, Alex has been trained in language since 1977, and by now his vocabulary includes more than ninety words. He can name objects ("What's this?" "green key") and count to six with a little better than 80 percent accuracy.

This work shows that even an animal with a brain as small as a parrot's can learn some aspects of language. This reinforces the lesson we learned in the last chapter: Complex behavior doesn't necessarily require complex neurological machinery.

As an aside, I should say that Alex's ability to count shouldn't surprise you. Hunters have known for generations that crows can count. The knowledge comes from observing that crows who see a hunter enter a blind won't come near it until the hunter leaves, and they'll do the same if two hunters enter the blind and one leaves. Only if three hunters enter the blind and two leave will the crows believe it's empty.

The ultimate test of language ability requires us to differentiate between humans and the great apes, and particularly between humans and chimpanzees. Unfortunately, the field of primate language acquisition went through a series of "Clever Hans" episodes in the 1970s and 1980s, incidents from which it still has not recovered.

The story begins in the 1940s, when two families of psychologists adopted baby chimpanzees and raised them with their own children. One of these chimpanzees, named Viki, eventually learned to say a few words (I can remember seeing a film of Viki saying "cup" when I took undergraduate psychology back in the mid-Pleistocene). The problem with this technique, of course, is that it requires the chimpanzee to make human sounds, a task for which its vocal tract is simply not suited. Thus, the experiment was not clean. The failure of the original chimpanzees to pick up language could be caused by something in their brains, but it could also be caused by the shape of their mouths.

The next attempt at teaching language to great apes started in the late 1960s, and centered around American Sign Language (ASL). ASL is not, as some people think, simply a vocabulary of hand positions. It is, in fact, a separate language with its own syntax and grammar, and, like all other human languages, it is in keeping with the deep rules that are built into our brains.

In the case of Washoe (a chimpanzee), Koko (a gorilla), and Nim Chimsky (a chimpanzee), extravagant claims were made about their ability to converse in this nonphonetic language. The apes appeared in all sorts of newspapers, magazines, and TV shows. They were, in fact, probably better known in their time than Clever Hans was in his. Unfortunately, as scientists began to examine the claims more closely, the story began to bear an unhappy resemblance to the Clever Hans saga. It appeared that the proponents of language acquisition had been overly generous in their interpretation of Washoe's and Koko's abilities.

Let me just give you a couple of examples to make this point.

One way that Washoe's signing was documented was to have a set of observers record every word. One of the observers was deaf, a "native speaker" of ASL. His comment on the experience was

> All the hearing people turned in logs with long lists of signs. They always saw more signs than I did. . . . Maybe I missed something, but I don't think so. The hearing people were logging every movement the chimp made as a sign.

In a similar incident, when the renowned chimpanzee ethologist Jane Goodall visited the laboratory where Nim Chimsky lived, she said that every sign Nim made was used by chimpanzees in the wild. Apparently the chimpanzee's natural repertoire of gestures was being interpreted by the researchers as ASL.

Sue Savage-Rumbaugh, whose work we'll describe in a moment, tells about her own experience with primatologist Roger Fouts and Washoe:

> [Roger] turned to Washoe, looked across the island, and noticed that a long rope lay near the center. . . . Roger turned to Washoe and signed, "Washoe, go get string there." He gestured in the direction of the string. Washoe looked puzzled, but did begin walking in the direction that Roger had pointed. She looked at a variety of things on the island, touching them and looking back at Roger, as if trying to determine what he meant. She walked past the string several times and each time Roger signed, "There, there, there (again pointing), there string." Finally, as she again approached the area where the string lay on the ground, Roger began to sign "yes, yes, yes" and nod his head emphatically. As Washoe reached the spot, she picked up the piece of string and was praised fulsomely. "See," said Roger, "she just had trouble finding the string." I was not convinced.

The consensus among scientists these days seems to be that

those early claims for language abilities in great apes cannot be substantiated.

So where does that leave us? Today, there is only one claim for language ability being advanced, and that is for a bonobo chimpanzee named Kanzi. (A very readable account of this claim is given in *Kanzi: The Ape at the Brink of the Human Mind*, by Sue Savage-Rumbaugh and Roger Lewin, published by Wiley in 1994.) Bonobo chimpanzees (*Pan paniscus*) are a different species from ordinary chimpanzees (*Pan troglodytes*), and they are sometimes called pygmy chimpanzees to make that distinction. They live in the rainforest in Zaire, south and east of the Congo (Zaire) River. They have a rather different sort of social structure than common chimpanzees, with more interactions and sexual encounters. The general folklore among primatologists since the discovery of this species in the 1920s is that they are the smartest of the great apes.

The Kanzi story begins in 1981 at the Language Research Center in Atlanta. Sue Savage-Rumbaugh was trying to teach Kanzi's adopted mother, another bonobo named Matata, to use a keyboard to communicate. The keyboard, about the size of a large serving platter, had symbols on each key. Thus, to "talk," all the chimpanzee had to do was to press the keys in sequence. Matata, who lived in the wild for her first five years, never really learned to use the keyboard. During the long training sessions, however, Kanzi was allowed to wander around the room, much as a human child would. To everyone's amazement, when it came Kanzi's turn to sit down at the keyboard, he already knew how to use it. He had, in fact, learned the symbolic language of the keyboard (as well as some spoken language) in much the same way a human child would have—more or less by osmosis.

Given the "Clever Hans" legacy, Savage-Rumbaugh and her colleagues have been very careful in designing their experiments. In one film she presented at a scientific meeting, for example, Savage-Rumbaugh showed Kanzi being tested on his ability to understand novel English sentences. She wore a welder's mask, so that Kanzi couldn't see her face, and sat stock still, so there was no body language. After asking Kanzi to pick up a ball and a bottle of soap, she said "Put the soap on the ball," a sentence Kanzi had never heard before. At this point, Kanzi picked up the soap bottle and squirted soap on the ball.

I have to admit that I find the evidence of Kanzi's linguistic abilities pretty convincing (although, as you might imagine given the history, there are many skeptical voices about this work in the scientific community). It does not seem to me that Savage-Rumbaugh is making particularly extravagant claims. The experiments show, she says, that Kanzi has about the same linguistic ability as a two-and-a-half-year-old child. One piece of evidence that I find particularly persuasive is that at one point in their work, Kanzi's trainers found that they had to start spelling words to keep him from understanding them—a procedure that any parent will appreciate.

If we take the claims being advanced for Kanzi at face value, where are we? We have a member of the most intelligent primate species, a veritable Shakespeare of nonhuman animals, raised under special and unusual conditions, performing at the level of a human child of two and a half years. But remember that in humans, real language ability starts just after this age. If the "grammar circuits" in our brains don't kick in until age three or so, as the evidence seems to indicate, then we have to conclude that even in this optimal case, animals other than humans cannot learn real human language.

There is more evidence for this conclusion. In the years since those initial findings, the length of Kanzi's sentences has not grown past a couple of words, nor has he displayed any of the types of development characteristic of innate grammar discussed above. Based on this finding, it seems safe to say that human language, as it is now understood, can be numbered among the unique adaptations of our species, and one that is not shared with any of the rest of the animal kingdom.

5

The Brain

Men ought to know that from nothing but the brain comes joys, delights, laughter and sports, and sorrows, griefs, despondency, and lamentations.

—HIPPOCRATES, ON THE SACRED DISEASE

Before we start getting into the details of the structure and functioning of the human brain, I'd like you to do a few things to get a sense of what an amazing organ it really is.

First, just close your eyes for a moment and then open them again. In a time too short for you to be aware of it, billions of cells in your brain took signals generated by light falling on the retina and reconstructed the visual field. *This is amazing!* As we shall see later, this simple process involves cells in many different parts of the brain working together (in ways we still don't fully understand) to produce the everyday experience of seeing with an efficiency far beyond the capability of any computer available today.

Next, close your eyes and think about an episode in your life that has a high emotional content—a time when you were very happy or very sad or very excited. A picture appears in your mind of a place and time far from where you are now, perhaps involving buildings that are no longer standing or people who are no longer alive. You may not have thought about this event for years, but the cells in your brain stored the image (and perhaps some of the emotion) and were able to replay them on command. *This is amazing!*

If you watch a brain growing in a fetus, you will see individual cells sending out tendrils to form the connections to other cells. Often the tendrils grow to a particular area and arrive before their intended targets are present. Like good hockey players, the developing cells move to where the "puck" is going to be, not where it is. *This is amazing!*

So when we conclude that what makes humans unique, what separates us from all other living things on our planet, has to do with the functioning of our brains, we are talking about an organ that is capable of achieving almost unbelievable levels of performance. In the next two chapters, we will go through the way the brain is built and how it works, starting with its basic building block, the neuron, and working up to our current understanding of how the parts come together to produce mental functioning. Before we get into the details, though, let me summarize here a few of the main features of the human brain.

1. Signals travel along a single neuron by a complex chemical process and are communicated to other neurons by the emission and reception of specialized molecules. They are *not* ordinary electrical currents.

2. Neurons in the brain are highly interconnected. They group together in spherical conglomerations called *nuclei* and in flat sheets called *cortices* (singular: cortex), each of which performs a highly specialized task. The whole assemblage is more analogous to a collection of semi-

autonomous villages than to a single, well-organized machine.

3. What we are and how we feel depends on the way molecules combine in the brain. The new understanding we have of the chemical functioning of the brain is causing a revolution in our treatment of mental illness. Antidepressant drugs like Prozac are, in fact, one of the first fruits of this knowledge.

4. We are just beginning to be able to map out the functioning of the various areas of the brain (and sometimes even of individual neurons) and understand how the whole system works.

The Chemical Factory That Makes Us Conscious

Like every other organ in the body, the brain is composed of cells. The main business of all cells is running chemical reactions, and the cells that are the main working part of the brain are no exception to this rule. Signals in the human nervous system are carried by neurons, but these signals are very different from things like electrical currents in wires or microchips. The first step in understanding the brain is understanding what neurons are and how they work.

The neuron, like every other cell in advanced life forms, has a complex internal structure that includes a nucleus (where DNA is stored), places where molecules from food are "burned" to provide energy, and places where various molecules important to the functioning of the cell are assembled. From our point of view, however, the most important events that take place in a neuron involve its outer membrane—the structure that separates the cell from its environment.

A typical neuron in the brain has a central body (think of this as a place that contains the machinery needed to keep the cell functioning) and a treelike structure leading away from it.

This treelike structure consists of a main trunk and a lot of branches, called *dendrites* (see Figure 1). Connections from different neurons in the brain are usually made to these dendrites, but can be made to other parts of the neuron as well. Think of the dendrites as the cell's primary input system. In addition, a long fiber called the *axon* leads away from the main body, devolving into branches that connect to different neurons. By processes we'll describe in a moment, nerve signals travel down the axon, through the branches, and make connections with other neurons. Think of the axon as the neuron's output system. Every neuron sends signals to, and is in turn signaled by, many other neurons—typically each neuron connects to a thousand or more others.

The neurons in the brain form a massively interconnected set of cells. To get a sense of how complex the system is, imagine yourself in a metropolitan area like that around New York City—an area that contains 10 million people. Then imagine taking a

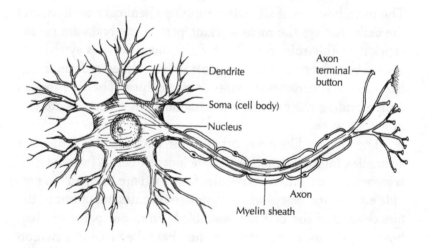

Figure 1. A neuron
Source: *The Sciences: An Integrated Approach* (New York: John Wiley & Sons, 1995), page 134.

(large) spool of thread and connecting yourself so that there was one thread running between you and each other person in the area. Then imagine everyone in the area doing the same. Can you even picture how much thread there would be, how connected each person would be to everyone else? The number of connections in the threaded city we've just imagined is roughly the same as the number of connections between neurons in your brain (although in the brain, as we shall see, the pattern of connections is different from what it is in this example).

The membrane of a neuron contains a number of different kinds of molecules, called *receptors*, sticking out into the surrounding medium on one side and into the interior of the cell on the other. Think of these receptors as icebergs floating in the cell membrane. The outer part of the iceberg is a convoluted molecule structure (think of it as a lock) that will fit only a specifically shaped molecule in the environment (think of it as a key). In fact, the carefully sculpted shapes of the receptors play many roles, including the following:

1. They act as the doors (or channels) through which, under certain conditions, atoms like sodium, potassium, and calcium can move into and out of the neuron.

2. They act as pumps—molecules whose shapes change in such a way that some atoms are moved from the outside into the cell, while other molecules are moved from the interior of the cell to the outside. For us, the most important of these pumps moves sodium ions (that is, sodium atoms that have lost one electron) to the outside of the cell and potassium ions to the inside. Sodium pumps play a crucial role in the propagation of nerve signals.

3. They act as receptors as described previously—molecules whose shape fits to the shape of molecules in the environment and which then initiate changes in the chemical operation of the cell.

When the nerve cell is not sending a signal (a state that physiologists refer to as *resting*), the channels that would allow sodium to enter the cell are mostly closed, while the potassium channels are mostly open. At the same time, the protein molecules that form the sodium-potassium pump operate to push sodium ions out of the cell and potassium ions into it. You can think of the way this molecular pump operates by picturing a posthole digger—one of those two-handled tools people use to dig cylindrical holes in the ground. When the digger is pushed down into the ground, it fits around a piece of dirt at the bottom of the hole. Energy, in the form of muscle power, is then applied to push the blades of the digger together and lift it and the trapped dirt out of the hole. In just the same way the pump molecule lodged in the axon membrane fits the sodium ion into it, then absorbs energy from another molecule in the cell and changes shape, expelling the sodium into the outside medium as it does so. On the reverse stroke of the pump, a potassium ion fits into the opened jaws on the outside and is pushed back in. The net result of this pumping is that the concentration of potassium ions inside the cell is higher than outside, while the concentration of sodium ions is higher outside than inside—think of the neuron as having fresh water inside and salt water outside. Because of this imbalance, the inside of the axon is negatively charged relative to the outside, and a voltage of about 70 millivolts (about 5 percent of the voltage across a typical AA battery) exists across the membrane of the axon.

When the axon is stimulated, a well-defined sequence of events takes place. The sodium channels open and positively charged sodium atoms move into the axon, pulled in by the negative charge that exists there. The sodium ions rush in until the charge inside the axon becomes momentarily positive, a state that distorts the molecules that make up the sodium channels and closes them again. Then, the change in charge opens more potassium channels, allowing positively charged potassium ions to flow out from the neuron, restoring the previous negative charge inside the axon.

The in-and-out flow of charge, with its sudden change of voltage, is called the *action potential*. As sodium ions flow into the axon, they spread out on the inside of the membrane, changing the charge on the upstream side of the signal and in effect causing the potential to move "down-axon." This process, in which sodium ions pour into the axon on the upstream side while potassium ions pour out on the downstream side, is what we call a *nerve signal*. After the disturbance passes a point in the axon, the pumps go back into operation and restore the resting state.

The action potential moves very slowly—generally no more than a fraction of an inch per second. In humans and other vertebrates, the axons are often sheathed with a substance called *myelin* that is impermeable to sodium and potassium. This sheathing has gaps in it, and its effect is to cause the action potential to skip from one gap to the next. This results in much faster propagation—signals can travel hundreds of yards per second (400 miles per hour) along a myelinated axon.

There are several important points to realize about the process I've just described. The most important is that it is nothing at all like electrical current running through a wire. Such a current is just a flow of loose electrons, with none of the complexity of the action potential.

Second, almost all of the detailed information about the way human neurons work was obtained by experiments on other animals, particularly the squid. The giant axon that runs down the body of the squid carries the signal for its "squeeze hard, squirt a lot of water, and get the hell out of here" response. The squid's axon is big enough so that scientists earlier in this century could insert their relatively bulky electrodes into it and measure voltages as signals went by. In fact, the mechanical and biochemical structure of neurons is pretty much the same across the animal kingdom—another illustration of the basic chemical identity of living things. A more recent example of this universality was the development of the first chemical test for Alzheimer's disease in 1994, a test that relied on studies of the mechanisms for memory in snail neurons.

Getting from One Neuron to the Next

The action potential moves down the axon and its branches until it gets to the end of the neuron. At this point, a new chemical process takes over to send the signal to downstream neurons. The end of one neuron doesn't actually touch the surface of another. Instead, there is a junction called a *synapse* connecting the two—a junction that includes a small gap over which the action potential cannot move. At the end of the axon branch of the neuron on which the action potential is traveling (called the *presynaptic neuron*) is a collection of little membrane sacs, called *vesicles*, each of which is filled with one of a specific collection of molecules. When the action potential arrives at the terminal of the presynaptic neuron, still other proteins in the neuron membrane are distorted so that they become channels for calcium ions. Calcium flows into the neuron, causing the vesicles to fuse with the neuron membrane and dump their molecular contents into the gap between the neurons. These molecules, called *neurotransmitters*, float across the gap and become the key that fits into the lock of receptors in the membrane of the downstream (or *postsynaptic*) neuron. When the neurotransmitters dock, they distort the membrane and produce signals that become part of the complex process discussed below by which the receiving neuron decides whether or not to initiate its own action potential.

Neurons often receive signals from a thousand or more other neurons. By some process that we haven't yet worked out, the neuron integrates these signals and then either initiates an action potential or doesn't. An analogy often used to describe the action of the neuron is to compare it to a rifle. A complex process goes into deciding whether the trigger is to be pulled, but once it is, the bullet proceeds according to its own set of rules—rules that are independent of the decision-making process. The rifle either fires or it doesn't. In the same way, a neuron either initiates an action potential (fires) or doesn't. If the neuron fires, however, the action potential moves according to the rules that govern the flow of sodium and potassium ions discussed previously.

Neurotransmitters, therefore, play a crucial role in the propagation of nerve impulses. There are many molecules that function as neurotransmitters, and different molecules have different effects on the postsynaptic neuron. Some tend to excite or promote the initiation of an action potential, others tend to dampen or inhibit that process. When some neurotransmitters dock, they distort the membrane around them and open or close ion channels directly. Others trigger chemical reactions inside the cell that also affect the voltage across the cell membrane, but do so more slowly. A few neurotransmitters can trigger both kinds of reactions, depending on the kind of receptor to which they attach. Finally, there is a class of small molecules called *neuropeptides* that can move across a specific synapse, but can also spread out and affect synapses at some distance from their point of emission.

Once the neurotransmitters have done their job at a particular synapse, they need to be cleared out so that the process can start again. The molecules may just diffuse, they may be broken down by enzymes devoted to that specific task, or they may be pumped back into the vesicles by a chain of molecular events analogous to the sodium-potassium pumping described earlier. This last process has been given the somewhat awkward name of *reuptake.* It's nature's way of recycling.

It's only within the last decade or so that medical researchers have started to understand and exploit the chemical signaling processes in the brain. The results have been profoundly revolutionary, both from a medical and a philosophical point of view. The point is that if you think of mental disease as something caused strictly by environmental factors (your relationship with your parents, for example), the kind of treatments you pursue will concentrate on those factors. The classical Freudian psychoanalyst, with a couch in his office, is a familiar example of this approach. If, on the other hand, you believe that mental disease is the result of brain chemistry gone awry, you are likely to look for ways of changing the workings of the brain's molecules instead.

One place where this new approach is being explored involves diseases we normally think of as "ordinary" medical conditions. Parkinson's disease, for example, results from a lack of a sufficient amount of a single neurotransmitter—dopamine—in the brain; migraine headaches can be alleviated by blocking a particular type of receptor for another neurotransmitter—serotonin. The most astonishing results, however, concern drugs like Prozac, which act to block the reuptake of serotonin at synapses. These drugs are powerful antidepressants, and because they act specifically on only one neurotransmitter, are relatively free of side effects. They are, I believe, striking examples of the wave of new chemical treatments for psychiatric disorders. Indeed, these so-called psychopharmaceuticals represent a major break with conventional psychiatric treatment, which focuses on techniques such as psychoanalysis and talk therapy. In its most extreme form, the case for the new look at psychiatry says that there is no point in lying on a couch and talking about your mother when you can get to the same result by taking a pill.

A small group of scientists has criticized the use of these sorts of drugs on the grounds that they treat only the symptoms of mental disease and not its cause. If you'll forgive me for climbing on a soapbox for a moment, I find these sorts of arguments hard to swallow. I have seen the effects of clinical depression on people close to me, and I have seen the changes in their lives when they started taking Prozac. The arguments of the critics of psychopharmaceuticals remind me of nothing so much as the story in the twelfth chapter of the Gospel of Matthew, when the Pharisees criticized Jesus for healing a lame man on the Sabbath. Who really cares whether a drug makes the world a perfect place, so long as it alleviates suffering?

But there is a deeper point here, one that is relevant to our discussion. The notion that the real cause of mental illness could not possibly lie in molecular activity in the brain is rooted in the dominant myth of the mid–twentieth century—the myth that every human being is a *tabula rasa*, affected only by his or her environment. What the success of drugs like Prozac teaches us is that this is simply not true. What we are and how we feel depends profoundly on chemical reactions in the brain. And that raises important questions about the nature of human identity. In the words of neurologist Richard Restak:

> What does it say about the human mind when one's general feeling of the world and one's place in it can be modified by a chemical that . . . acts so subtly that the person taking the drug has no side effects or other experiences usually associated with the taking of a medication?

What indeed!

The Structure of the Brain

The brain is not a random collection of neurons. To be sure, it contains many neurons—some 100 billion. (For reference, this is about the same as the number of stars in the Milky Way galaxy, and about 10 million times more than the number of stars you can see on a clear night.) But these neurons are not haphazardly arranged. The brain is a complex and highly structured organ.

The first point to make is that the brain is not just neurons. Like every other organ in the body, it is laced with blood vessels to bring oxygen and nutrients into its cells and carry wastes away. The blood also carries other molecules into the brain, a point to which we will return later. Furthermore, as many as 90 percent of the cells in the brain are not neurons but things called *glial cells*. They are generally smaller than the neurons, and it has been thought that they played a largely supportive role in the brain—think of them as surrounding and nurturing the neurons. Lately, however, there has been some suggestion that they may actually play a role in setting the threshold at which the neuron fires.

Many neurons in the brain are found in specific groups that perform specific functions. Some of these, roughly spherical in shape, are called nuclei, while others, in which neurons are arranged in layers, are called cortices. Nuclei and cortices make up what is called the "gray matter" in the brain. Axons from these structures tend to be arranged in bundles of fibers, with each axon covered with myelin. This is the "white matter" of the brain (myelin has a whitish color).

To get an idea of the overall structure of the brain, imagine putting on a couple of boxing gloves (see Figure 2). Now imagine crossing your hands over, so that the little fingers of your two hands are side by side. Finally, imagine that you are in a room with a tall, narrow pedestal whose top is a bulb that just fits into the palms of the gloves. Put the gloves on top of the bulb and remove your hands. The result gives us a useful way to picture the large-scale structure of the brain.

The tall pedestal is the spinal cord, carrying nerve impulses to and from the brain. The lower part of the bulb on the pedestal is a set of organs that are referred to as the *brain stem* and the *cerebellum*. This part of the brain is primarily concerned with regulating basic body functions. The cerebellum, for example, is involved in monitoring body positions and maintaining balance—reach out to pick something up and your cerebellum takes care of all the little motions of the muscles in your back and neck that keep you upright during the operation. Other parts of this section of the brain control functions such as breathing, heart rate, and vomiting.

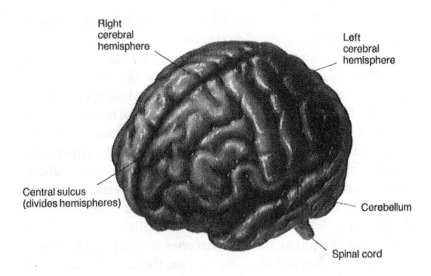

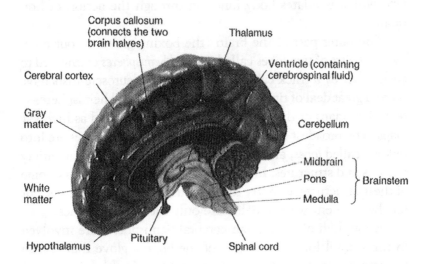

Figure 2. The structure of the brain
Source: *The Sciences: An Integrated Approach* (New York: John Wiley & Sons, 1995), page 607.

Just above the brain stem (the top of the knob in our analogy) is a region called the *diencephalon*, which serves as the major coordinating center of the brain. Here we find the *thalamus*, two egg-shaped bundles of neurons that serve as the main relay station for nerve signals between the brain stem and the upper reaches of the brain. Just below the thalamus we find the *hypothalamus*, a collection of neurons that is involved in activities associated with the sex drive, hunger, thirst, pleasure, and pain. The hypothalamus also has an intimate connection to the *pituitary gland*, which is the master gland of the endocrine system. Specialized neural cells in the hypothalamus produce small molecules that travel to the pituitary gland via a special system of veins, and once there they affect the production of hormones in the pituitary itself. Through this chemical signaling process, the brain is connected to the body's other great control mechanism, the *endocrine system*, which regulates body function through the action of hormones.

The outer part of the brain (the boxing gloves in our analogy) consists of two lobes called *cerebral hemispheres* connected to each other by a thick cable of nerve fibers. Neuroscientists have spent a great deal of time mapping out the cerebral hemispheres—technical maps of the brain are every bit as detailed as highway maps. The broadest divisions of the cerebral hemispheres are into regions called *lobes*, and each lobe is broken down into a variety of areas and structures. In the next chapter we will explore some of these structures as we try to work out the brain's functions, but for the moment we will delineate only the broad outlines.

About half of the human cerebral hemispheres are involved in the frontal lobes—the parts of the boxing glove that encase the fingers. Among other functions, neurons in this lobe control conscious movement. The rearmost parts of the brain, where you put your hands into the boxing gloves, are the *occipital lobes*. The word means "back of the head" in Latin. This is where the initial processing of vision takes place. Between the frontal and occipital lobes—the back part of the boxing gloves on each side—are the *parietal lobes*. The word means "wall" or "divider" in Latin. This is where information about the state of the body is processed. Finally, the thumbs of the boxing glove form the *temporal lobes*, which are involved in hearing, memory, learning, and emotion.

The outermost layer of the brain—what would be the leather in the boxing gloves—is highly wrinkled and about an eighth of an inch thick. This is the cerebral cortex. As we shall see in the next chapter, this is the region of the brain that is associated with what we call the higher mental faculties. It is linked to the diencephalon and the hypothalamus by a loop of neurons called the *limbic system*, which is involved both in the phenomenon of memory and such basic drives and emotions as hunger, thirst, and sexual arousal.

As we go from the spinal cord to the outer layers of the cortex, we move from the deepest and most visceral parts of our nature to the "higher" functions, from the most generalized to the most specialized. Unfortunately, this understanding of the brain has led to a rather oversimplified notion of brain function in some parts of the popular press—in which the brain is seen as a set of successive overlays. At the bottom (the brain stem and diencephalon) is a kind of primitive, reptilian brain shared with all animals, with progressive overlying refinements added until we get to the cerebral cortex, which reflects the highest brain functions. In its extreme form, this view presents an idea of the brain as a kind of sedimentary structure, like the stratifications of the Grand Canyon. Each new layer adds a new function, while underlying layers stay more or less the same.

This is another of those concepts that the French call a *fausse idée claire*. It's simple, elegant, clear, and completely wrong. In fact, most of the major parts of the brain are present in all vertebrates, and were presumably present in all our ancestors. The process of evolution has produced widely different brains by selective development of different parts of the basic system—by adding neurons to expand a particular part or by rearranging the neurons that are already there.

In addition, it's not so easy to make a functional separation of different parts of the brain. It's much better to think of the brain as a coherent system, with each part communicating with the others. Although distinct features and functions can be associated with specific groups of neurons, it is nevertheless true that these groups are always in contact with other groups, and that no part of the brain really acts in isolation. The brain can, in fact, be thought of as being composed of a large number of interacting groups of neurons, which, in modern terminology, makes it an example of a complex system. This is a theme to which we will return repeatedly throughout this book, for it is the key to the brain's function and to human uniqueness.

Growing Neurons

Every human being begins as a single fertilized cell, or *zygote*, in his or her mother's fallopian tube. After about three weeks, the embryo is about one-eighth of an inch long and looks something like a corn cob (the "kernels" will eventually develop into the spine). At the top of the corn cob are some structures that lead to a hollow region in the interior known as the *neural tube*. It is the cells in this neural tube that will eventually multiply to form both the brain and the rest of the central nervous system. At the end of four weeks, the cells near the top of the neural tube have grown to form a sinuous curved structure, and the top part of this curved structure is what will eventually develop into the brain. By eleven weeks, there is a clear bulge at the top of the fetal spinal column, and by five months the general features of the brain can be seen in rough outlines.

The general process by which the complex structure of the brain is elaborated is that cells will migrate to a particular area, and then mature and differentiate. In other words, the development of the brain in the fetus, as with many other organs in the body, proceeds by the construction of a rough outline first, followed by intense elaboration. If you have ever watched a large building under construction you have seen the same thing. First the steel frame goes up and the building is closed in. At this point, the rough outlines of the building can be seen. Nevertheless, it may take many months for the finish carpenters, electricians, plumbers, and other tradesmen to turn that rough shape into a finished building. In just the same way, the rough outlines of the brain can be seen early on in the fetus, but the elaboration of the structure takes many months.

Perhaps the most intriguing thing about the development of the brain is that synapses between neurons in the brain do not start to form until about the seventh month of development. (My colleague Harold Morowitz and I, in our book *The Facts of Life: Science and the Abortion Controversy*, have pointed out that this feature of brain development is not without implications for the bitter debate over abortion in the United States.) For our purposes, however, we shall simply note that the most important aspect of brain structure—its interconnectedness—happens fairly late in the development of the fetus. The process by which the brain makes the connections between neurons by forming synapses illustrates a point that I have made repeatedly in the previous chapters—that the brain is a chemical system whose function depends on the shape of particular molecules.

If you think about it for a moment, you will realize that the process by which an individual neuron chooses neurons with which to form synapses must be extremely complicated. In fact, the brain begins with about twice as many neurons as will eventually be present. As each neuron begins to grow axons and dendrites, the growth of these structures is determined by chemical signals in the environment. In essence, the axon branches act like bloodhounds tracing their way to their target by following specific molecular signals. In fact, the axons often arrive at their final location before the target neuron is functioning, a feat which recalls the comment of hockey great Wayne Gretzky, "I don't skate to where the puck is, I skate to where it will be." If the neuron doesn't make the right connections, it "commits suicide" and disappears. The process by which a cell knows that it is supposed to do this is also chemical, and understanding it in detail remains one of the major areas of research in molecular biology.

The important point to take home is that the brain is not "designed" from scratch. Instead, the brain grows and forms synapses according to well-defined chemical signals. It's not a very efficient process, because half of the neurons that begin making connections wind up dying off.

And although your brain went through a period of intense development when you were *in utero*, it has never really stopped changing. The sentence you've just read, for example, is now in your short-term memory, and it certainly wasn't there a minute ago. If you wanted to, you could memorize the sentence so that you could recite it years from now.* This means that the synapses in your brain are constantly in the process of being strengthened and weakened. Your brain never stops developing and changing. It's been doing it from the time you were an embryo, and will keep on doing it all your life. And this ability, perhaps, represents its greatest strength.

---

*Please don't—there are an infinity of sentences that would better repay memorization.

6

Of Tamping Rods and Grandmother Cells
How the Brain Works

For in and out, above, about, below
'Tis nothing but a magic shadow show

—The RUBÁIYÁT OF OMAR KHAYYÁM,
transl. by Edward FitzGerald

An Accident and Its Aftermath

That day in the summer of 1848 was just another workday for Phineas Gage. He was a foreman of a blasting crew putting in a new railroad line near the town of Cavendish in western Vermont. In those days, men would cut holes in rocks with long, pointed steel augers and sledgehammers, then put black powder into the holes. Before the powder could be ignited, it had to be tamped—

79

pounded down firmly with a long steel rod. Tamping was Phineas Gage's job. It involved taking one of the augers, reversing it so that the blunt end was down, and pounding on the black powder in the hole. This was a normal operation, done dozens of times a day, but on that particular day something went wrong. No one knows why—maybe the rod struck a spark on the rock as it was pushed down. Whatever the reason, the powder exploded, blowing the pointed steel rod back out of the hole. It caught Gage on the left side of the face, entering just under the cheekbone. It traveled through his brain, exiting near the center of the skull.

Miraculously, despite the fact that a 3-foot steel rod had been shot through his head, Gage survived. In fact, except for a short period of unconsciousness, he was awake, alert, and capable of talking to his friends as they rushed him to town and got a doctor. He was soon up and about, but people noted a strange change in his behavior. Before the accident, Gage had been a steady, reliable kind of man—indeed, it was this responsible behavior that had gotten him the job as a foreman in the first place. After the accident, he seemed unable to make long-range plans. He started drinking and swearing (neither of which he had ever done before) and seemed unable to control his temper. He lost his job and starting drifting from place to place, occasionally working in circus sideshows (where he would be on display next to the steel rod that had caused his injuries). He died in San Francisco in 1861. Because the Civil War was then raging, physicians on the East Coast who had followed his case were unaware of his death, and there is no record of an autopsy being done. After the war, Dr. John Harlow, who had first treated Gage, contacted Gage's family and persuaded them to allow the body to be exhumed and the skull returned to Warren Museum of the Harvard Medical School.

In 1992, neurophysiologist Hanna Damasio and her co-workers took careful measurements of the holes in Gage's skull. Using modern techniques of computer imaging, they were able to make a pretty good guess about the parts of his brain that were affected by the passage of the tamping bar. Their calculations indicated that the bar passed through a part of the brain known as the *ventromedial prefrontal region*, located on the lower forward section of the frontal lobe (see Figure 3). Studies of other people who have sustained damage to this region (because of a tumor or a stroke, for example) indicate that they seem to exhibit the same sort of behavior change as that recorded for Phineas Gage. They seem unable to understand the need for long-range planning, and thus to act irresponsibly.

The story of Phineas Gage, as unfortunate as it is, nevertheless represents one path that scientists have followed to learn about the functioning of the brain. By accident or disease, someone suffers the loss of a specific section of the brain (the specific na-

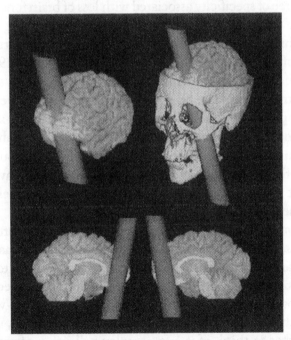

Figure 3. A computer reconstruction of the areas of the brain of Phineas Gage destroyed by his accident
Source: Courtesy of Hanna Damasio. As published in A. Damasio's *Descartes' Error* (New York: Grosset/Putnam, 1994), page 32.

ture of the damage is often not known precisely until autopsy). That person's mental functioning is then observed, usually as part of the ongoing treatment of the condition. Over the years, enough knowledge gleaned from cases of this sort has given us a pretty good notion of the large-scale functioning of the brain.

The other path to this sort of knowledge, one that we'll use extensively later in this chapter, involves experiments done on animals. The closer the animal subject is to *Homo sapiens*, the more confident we can be of the extrapolation involved in interpreting the data. Most of the knowledge we have of the neural basis of vision, for example, comes from work on cats and monkeys.

As the case of Phineas Gage illustrates, there is an astonishing degree of specificity associated with loss of brain function. The damage to his brain did not affect his eyesight, his language ability, or his coordination—just his behavior. Another striking example of this sort of specificity occurred in Montreal in 1953, when a young factory worker called H. M. underwent brain surgery in an attempt to alleviate his epilepsy. The surgery removed parts of the temporal lobes, and, although he had fewer seizures afterward, his memory was severely affected. He was able to remember clearly everything that happened to him up until the surgery but remembers nothing that has happened since. Doctors who have treated him for years, for example, have to introduce themselves each time they meet him.

From these sad stories, and many like them, an important truth emerges. The brain is not like a giant corporation made up of interchangeable general-purpose parts. Instead, it seems to be more like a collection of little villages, each performing a specific task, each connected to other villages and fitting into the whole. In fact, rather than thinking of the brain as a single organ, it is probably better to think of it as a complex collection of organs. Just as the digestive system has the stomach, liver, intestines, and so on, the brain has many different parts that must work together.

Another useful analogy to illustrate this aspect of the brain is to think of it as something like an orchestra. Each instrument does its own thing, but the net result is a symphony.

Lewis and Clark in the Brain

Working out the details of how the brain functions is one of the great ongoing tasks of science. There is, I think, an analogy between this scientific exploration of the inside of the human skull and the process by which Europeans explored North America. First there were activities like the Lewis and Clark expedition—explorations whose purpose was finding the general outlines of the new terrain. They were followed, in due course, by the National Geological Survey, which was charged with providing extremely detailed maps of the new terrain.

The stories about brain-damaged individuals like Phineas Gage illustrate, I think, Lewis and Clark expeditions into the brain. The purpose of this sort of work was (and remains) an attempt to find which brain area or areas are involved in each sort of mental activity. This part of the exploration of the brain, though well begun, still has a long way to go, as we shall see. But even before it is done, the geological survey of the brain is beginning. This work attempts to describe the brain at the level of the individual neurons, rather than large brain areas. Later we will talk about the process of vision to illustrate this approach and to try to give some sense of how far it has gone and how far it has yet to go.

We can illustrate the kind of broad-gauge, Lewis and Clark maps of the brain with a few examples. Put the thumb and forefinger of your right hand on the two sides of your right ear, then move your hand across your skull until the fingers are at the corresponding places on your left ear. In this motion, you have traced out two very important regions of the cortex—the *primary motor cortex*, located along the back of the frontal lobe, and the *primary somatosensory cortex*, located along the front edge of the parietal lobe. As the names imply, these two strips of cortex control

primary motor cortex   **primary sensory cortex**

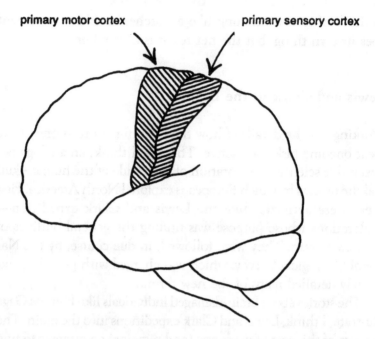

Figure 4. The primary motor cortex controls body movements. The primary somatosensory cortex receives sensations from different parts of the body.

movement of and receive sensation from different parts of the body. The motor cortex in the right hemisphere controls movement on the left side of the body and vice versa. Starting down in the cleft between the hemispheres in the middle of the brain and working up to the top of the hemisphere and then around to the region over the ear are neurons that control (or receive signals from) different regions of the body. If you feel something with your left big toe, it is neurons in the right somatosensory cortex down in the cleft between hemispheres that are firing. If you move your toe, it is neurons in roughly the same position in the primary motor cortex that give the command. Moving around the hemisphere, we find neurons associated with the legs and trunk at a position of about twelve o'clock, those connected with the arms at about

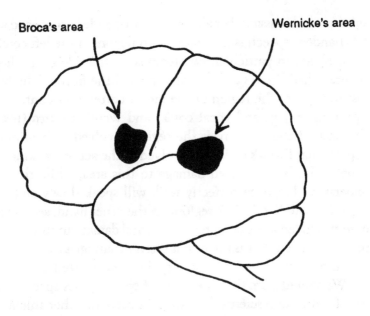

Figure 5. Broca's area and Wernicke's area in the brain

one o'clock, those connected with the hand at about two o'clock, and those connected with the mouth and jaws at about three o'clock.

The frontal lobe forward of the primary motor cortex is primarily devoted to the processing of neural signals and to what we usually think of as the higher mental functions. It is the area of the brain that is most highly developed in humans compared to other animals. It is, in fact, the existence of the frontal lobe that gives the human forehead its characteristic bulge. The connection of this region of the brain with the characteristics we usually lump under the term "intelligence" is reflected in slang terms like "highbrow."

A detailed description like the one above for the primary motor cortex could be given of the function of different regions of the frontal and other lobes. For the moment, let me just mention two other areas of the brain that will be important in the subsequent discussion. For almost all right-handed people and most left-handers, speech is associated with regions of the left cerebral hemisphere, in particular, two regions on the side of the hemisphere called *Broca's area* (located toward the front of the head, just in front of the region of the motor cortex that controls the lips, tongue, jaw, and vocal cords) and *Wernicke's area* (toward the back of the head, near the region involved in hearing). It appears that Broca's region is involved in the act of speaking, and people who have suffered damage to this area, while they can understand language perfectly well, will speak slowly and haltingly, if at all. Wernicke's region, on the other hand, seems to be involved in language comprehension, and damage to this area will result in speech that is fluent but often meaningless, as well as in impairment of understanding of spoken and written language.

We should note that the study of speech poses special problems for the brain scientist. As we have seen, no other animal has developed human speech. Thus, there are no meaningful animal experiments that could cast light on speech functions of the human brain.

The presence of speech areas illustrates the notion of the brain as a collection of "villages," or, to use the technical term, the brain's use of *distributed processing*. If you want to say something, you first have to form the thought somewhere in the frontal lobe, then send a signal back through Broca's area, and from there to the primary motor cortex to move your lips, tongue, and vocal cords.

Monitoring the Living Brain

Since the mid-1980s, two important new tools for exploring the functioning of the brain have become available to scientists. Both of them have the enormous advantage of allowing them to watch the functioning of the human brain in a completely nonintrusive way. Both were originally developed as diagnostic tools to be used in neurology, but once they had been created it quickly became evident that they could contribute enormously to our understanding of the nature and operation of the brain. In order of appearance, these techniques are *positron-emission tomography (PET)* and *functional magnetic resonance imaging (fMRI)*.

Both techniques depend on the way neurons, like every other cell in the body, get their energy from molecules carried in the blood. When a cell is carrying out its function, whether it is a muscle cell contracting or a neuron firing, it requires more energy than when it is resting. The body meets this need by increasing the flow of blood and nutrients to those cells—this is why your heart rate goes up when you exercise. Bodybuilders make use of this situation when they compete. Before going on stage, they "pump up" their muscles by lifting weights. The muscles become engorged with blood and look better for the judging.

In just the same way, when neurons in your brain fire, they "pump up." The flow of blood to the active region increases, and although the increase in the blood flow is small compared to that seen in muscles, it is nevertheless real and measurable. PET and fMRI are two different techniques for measuring the increased blood flow to regions of the brain that are being used.

PET scans require the use of a radioactive isotope of oxygen, oxygen-15. This oxygen molecule has to be prepared in a special nuclear facility, and once prepared it shares the central property of all radioactive materials—it has the same chemical reactions as every other oxygen molecule, even though the nucleus of the atom will disintegrate after a while. The oxygen-15 is attached to or incorporated into another molecule—water, for example, or glucose—which is then injected into the bloodstream. In a matter of tens of minutes, the oxygen-15 nucleus will disintegrate, spitting out fast-moving debris that includes a particle called the *positron*. The positron is an example of antimatter. It has the same mass as an ordinary electron, but a positive electrical charge. When a positron encounters an electron, as it must shortly do after it is emitted by the oxygen, the two undergo a process called annihi-lation.* The positron and electron disappear and their energy appears as a pair of extremely energetic X rays. These can be detected outside the body, and a computer can put together data from many such annihilations to produce a three-dimensional pic-ture of the location of oxygen-15 atoms (and the molecules of which they are part) in the brain. (This sort of computerized production of a picture from data is called *tomography*, which explains the *T* in PET.)

The point about the PET scan, of course, is that it can moni-tor brain activity while that activity is taking place. When it first became available in the 1980s, the scientific literature suddenly blossomed with color photographs of brain cross sections with different parts lit up to illustrate the effects of different mental

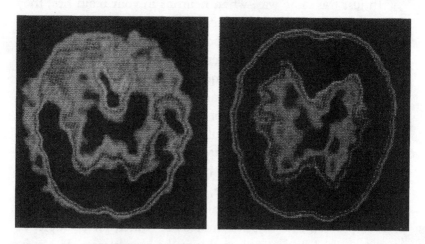

Figure 6. A PET scan
Source: *The Sciences: An Integrated Approach* (New York: John Wiley & Sons, 1995), page 307.

---

*Annihilation will be familiar to aficionados of *Star Trek*. It's what happens when material from the matter and antimatter pods comes together in the di-lithium crystal to produce warp speeds.

activity. People quickly saw that different parts of the brain were involved, for example, in thinking about a word, thinking about saying a word, and actually saying the word. It seemed as if a fundamental barrier to the understanding of the human brain had been removed.

In addition to providing information about brain function, PET scans can produce data that no other techniques can. For example, there are plans to use them to map out the locations of receptors in the brain by incorporating radioactive atoms into various neurotransmitters. There is even a possibility that they may allow us to trace pathways along which nerve impulses travel.

There are some limitations to the technique, however. For one thing, it requires the capability of creating and handling radioactive materials—not the sort of thing you'd find in the average psychology laboratory. For another, it takes some time for the image to form—perhaps up to a minute. This means that it's hard to catch swiftly moving events in the brain. On the other hand, the technique is enormously versatile.

If I was agreeably surprised to see those first PET scans of functioning brains, I was stunned when data from fMRI machines began to become available. Magnetic resonance imaging* is a technique that depends on the properties of the atomic nucleus, particularly the nucleus of the hydrogen atom, a particle called the *proton*. Like the Earth, the proton spins around an axis and has a north and south magnetic pole. If a proton finds itself in a magnetic field, its magnetic axis starts to describe a lazy circle in space. You can see this effect, called *precession*, in a common child's top. While spinning rapidly around its axis, the top can also move so that the axis itself describes a slow circle.

The rate of precession for a proton in a magnetic field depends on the strength of that field. If the area around the proton

---

*Formerly *nuclear* magnetic resonance imaging, but "nuclear" was dropped to ease public fears during the 1970s.

is flooded with radio frequency waves, the waves that have the same frequency as the proton's precession will be absorbed and emitted in a predictable way. By monitoring the strength of the radio frequency waves, then, we can make extremely precise measurements of the proton's precession and hence of the magnetic field in which it finds itself.

Ordinary MRI imaging—the sort of procedure you'll find in almost any hospital these days—uses these sorts of measurements to estimate the number of protons in different regions of the body and hence to differentiate between different types of tissue. This is how it produces such amazingly sharp and detailed images of the body's interior. The common medical use of MRI in this mode has begun to be called *structural* MRI, or sMRI, to distinguish it from functional MRI.

On the other hand, fMRI uses extremely precise measurements of proton precession to measure small changes in the magnetic field at the location of the proton. Blood is slightly magnetic, so a small change in flow produces small changes in the magnetic field in the region around the capillaries, and it is these small changes that are picked up by fMRI machines.

In 1994, some research using the fMRI technique caused a minor flurry in the nation's media. Scientists at Yale had studied the brains of men and women engaged in various language tasks. They found that although men and women seem to speak the same language, their brains produce that language quite differently. Men's speech tends to be mostly the result of neurons firing in the left hemisphere, while women's seems to come from areas on both sides of the brain. Of course, as many wags pointed out at the time, any married person could have told you that while men and women use the same words, they don't really speak the same language.

Many fMRI scans have turned up other fascinating things about the way the brain works, providing a kind of scientific underpinning for common folklore. For example, scans taken when people are set the task of memorizing a face show that the older a person is, the less blood flow there is to the regions of the brain where memories are encoded. This settles an old scientific controversy about why it gets harder to remember things like telephone numbers as we age. There used to be two schools of thought on this—one held that memories weren't formed as easily, the other held that the memories were formed, but that the retrieval process got worse with age. The fMRI data now seem to support the former view.*

The primary limitation of the fMRI technique is that, unlike PET, it is tied to blood flow. This means that the smallest details of brain function it can resolve are those associated with circulatory watersheds in the brain. The capillaries in the brain are arranged so that if neuron A requires more blood, all the neurons within about a square millimeter of A also get increased flow. This situation sets a fundamental limit to the resolution we can obtain with this technique. As of this writing (fall 1996), in fact, fMRI machines were already capable of reaching that limit.

At the moment, the primary use of both PET scans and fMRI machines is in clinical medicine. It seems that many mental diseases have characteristic patterns of brain activity. People suffering from obsessive-compulsive disorder, for example, show abnormal patterns in three distinct regions of the cortex and regions just beneath it. In fact, the usefulness of the instruments for diagnosis and treatment of patients poses problems for basic researchers, who have difficulty getting time on the machines.

Nevertheless, it is clear that the availability of these kinds of instruments will, over time, allow us to produce extremely detailed maps of the way the brain functions. Walter Schneider of the University of Pittsburgh, for example, speaks of mapping out the functions of the cerebral cortex, square millimeter by square millimeter, as one of the great scientific adventures of the coming

---

*I am reminded of a comment attributed to Albert Einstein: "The three worst things about getting old are that you lose your memory, . . . that you lose your memory, . . . and I forget the third." We should all age as gracefully as he did!

decade—an adventure fully as engrossing and exciting as the sequencing of the human genome. (For reference, a square made from a three-by-three array of dots the size of the period at the end of this sentence is about a square millimeter.)

Seeing Is Believing

However, even a map of the cerebral cortex on a scale of square millimeters doesn't get us to our goal of understanding how the brain works in terms of the functioning of neurons. To get to that level, we have to go to experiments in which the activities of individual neurons can be monitored. The technology exists to insert microprobes into individual axons and record the rate of firing on the neuron as specific mental tasks are performed. For obvious reasons, this sort of experiment can only be done on nonhuman animals, but our brains are sufficiently like those of other primates that information acquired in this way can often be applied directly to human beings.

One of the pioneering experiments of this type was done by Patricia Goldman-Rakic at Yale University. Her group monitored the activity of single neurons in specific areas of the cortex in the frontal lobes of monkeys as they performed tasks involving short-term memory. In essence, the monkey had to remember where a flash of light in its peripheral vision field was for a short time, then move its eyes to that spot. In what has to be one of the most dramatic demonstrations of the connection of individual neurons to a specific mental activity, the group was able to see the neurons turn on when the monkey was remembering the position of the spot, then turn off when the monkey looked at where the spot had been. They were even able to trace the firing of neurons from the frontal lobe to other parts of the brain as the monkey turned to look at the spot and the neurons in the frontal lobe returned to normal. This is a classic example of understanding brain function at the level of neurons.

As it happens, the brain system that can be understood best at this level is the process by which patterns of incoming light are turned into mental images; although, as we shall see, even here our knowledge is most definitely limited. The first step in the process, the conversion of incoming light to neural impulses, is carried out by cells called *rods* and *cones* in the retina of the eye. (The names of the cells come from their shape as seen in a microscope.) In these cells, a complicated chemical process turns the energy of the incoming photons into action potentials.

The retina of the eye, however, is not like a piece of film that simply passes on the information it receives. The output from many rods and cones in a particular area is relayed to another set of cells in the retina called *ganglion cells*. One type of ganglion cell will fire if the signals it is getting correspond to a bright spot with a dark "surround," while another type will fire only if there is a dark spot with a bright surround. Thus, the information in the signals that leave the retina has already been processed.

In passing, I should remark that physiologists have long classified the retina as part of the brain, rather than as part of the eye. This is because if you watch a fetus develop, the retina grows out of the same cells that produce the brain and the rest of the central nervous system. The fact that the retina is where visual information processing begins means that it functions as part of the brain as well.

There are actually three different kinds of ganglion cells in the retina, each responding to a different aspect of the incoming light. Some respond to colors, while others respond to small differences in intensity. This means that even in the retina, several different ways of processing the incoming information are going on at once.

When they enter the brain proper, the axons of most ganglion cells connect to a cluster of neurons in the thalamus called the *lateral geniculate nucleus*, or LGN. In the language of the neurophysiologist, we say the ganglion cells "project" onto the LGN. In addition, some ganglion cells project onto a cluster of neurons at the top of the brain stem called the *superior colliculus*. We'll discuss the difference between these two projections in a moment.

The neurons that project to the superior colliculus seem to produce a rough, coarse-grained picture of the visual field. They do not include any ganglion cells that respond to color, for example. The purpose of these neural signals seems to be to give early indication of movement, particularly in the periphery of the visual field. When neurons in a particular region of the superior colliculus fire, they seem to initiate an automatic response that brings the area of motion into the center of the field of vision. You have probably had this experience. Maybe you were standing talking to someone in a room when something unexpected happened—someone came through a door, for example, or a window curtain billowed. Both you and the person to whom you were talking immediately turned and looked in the direction of motion. That everyday event was caused by the neurons in this area of the brain.

Most of the ganglion cells, however, project to the LGN, which are two little bumps on top of the thalamus. LGN on each side of the brain receive signals from both eyes, so that the left side of the brain processes input from the right side of the visual field, and the right side of the brain processes signals from the left side of the field. Ganglion cells near each other in the retina project onto nearby cells in the LGN, so that there is a rough map of the visual field in the LGN neurons. The main function of the LGN seems to be as a kind of relay station, taking the signals coming in from the retina and sending new signals to a region called the *visual cortex*, at the back of the occipital lobe.

Put your hand on the back of your head. The lump you feel is the skull over the visual cortex of your brain. The cortex in this particular region resembles a rather messy layer cake, with different-shaped neurons predominating in different layers, but with all the layers connected to each other by axons and synapses. Physiologists distinguish six such layers, numbering the outermost layer 1 and the innermost 6. The different layers appear to perform different functions in the analysis of visual data.

Axons from the LGN feed mainly into layer four of the visual cortex, so you can think of this as the "input" layer. After having been processed as described below, signals are sent from the visual cortex to other parts of the brain. Depending on where the signals are going, they leave from different layers. For example, signals to other parts of the cortex project out mainly from layers two and three, while those to noncortical parts of the brain project out from layer five. In addition, some of the neurons in layer six project back to the thalamus (the purpose of these back-traveling signals is not understood).

It is in the layers of the visual cortex that the picture that had been broken up into nerve signals in the retina begins to be reassembled. The general strategy is that specific neurons fire only when a specific feature, represented by the light and dark dots relayed from the ganglion cells, is present in the visual field. For example, there are neurons that will fire only when a horizontal edge appears, others that fire for a vertical edge, still others for edges tilted at a specific angle. Each of these neurons takes in data from many ganglion cells but fires only if that data corresponds to the specific feature for which it seems to be programmed. As was the case in the retina, all of these processes are going on simultaneously—neurons are firing in response to a horizontal edge in one part of the visual field at the same time as other neurons are firing in response to a vertical edge somewhere else. In the jargon of computer scientists, this is called "parallel processing."

Neurons from the visual cortex project out to other regions on the parietal and temporal cortices—a good part of the rear section of the cerebral cortex is devoted to visual processing. As the process of reassembling the picture continues, we know that there are cells that fire only when more complex shapes appear in the visual fields—star shapes, for example, or circles with bars through them. At this point, however, our detailed knowledge of what the neurons are doing begins to give out. We don't know how the brain assembles these preliminary building blocks into the coherent visual picture we experience.

Cognitive scientists often speak of this problem as an attempt to understand how to pull together, or "bind," the various threads of reconstruction of the visual image that we know are moving forward in the brain. This so-called binding problem remains one of the brain's great unsolved mysteries.

There are, of course, theories about how binding might take place. For a while, for example, some people took an almost Platonic position and argued that the brain was arranged in a kind of hierarchy. Just as "edge" neurons will fire only when they get input from specified ganglion cells, it was suggested, there are secondary neurons that will fire only when they receive input from a certain set of "edge" neurons, then higher cells that fire only when they receive signals from certain sets of secondary neurons, and so forth. The idea was that there was an upward-moving cascade of neuronal activity, culminating in the firing of a few neurons whose signals (in some unspecified way) triggered the sensation of seeing something. In its extreme form, this concept was encapsulated in the notion of the "grandmother cell"—the one cell in your brain that would fire when you saw grandma.

This idea was rejected for many reasons, not the least of which is that there are not enough cells in the brain to represent all possible visual fields. You couldn't, for example, have just one grandmother cell. There would have to be cells for grandma in a red dress, grandma in a blue dress, grandma smiling, grandma frowning, grandma ten feet away, grandma five feet away, grandma riding her Harley-Davidson, and so on. Furthermore, it is quite possible to call up mental images—a rhinoceros in a football uniform, for example—for which it is clearly ridiculous to suggest a reserved cell somewhere in your cortex.

More recently, scientists have suggested that the final product of the neural cascade that we have traced from the ganglion cells to the visual cortex and beyond is not the firing of a specific cell but a patterned firing of many neurons. The idea is that the product of the process of seeing is not a single neuron firing, but a lot of neurons firing in a specific pattern. You can visualize this kind of coordinated firing by picturing neural signals sloshing back and forth across a region of your cortex like water sloshing back and forth in your bathtub. In this scheme, each visual image corresponds to a different pattern of "sloshing," and individual neurons may participate in producing many different visual experiences. Not only does this suggestion get around the numerical difficulties encountered by the grandmother cell hypothesis, but scientists are beginning to find evidence for these sorts of oscillations in the brain. It appears that groups of neurons fire in concert at a basic rate of forty times per second, and some scientists have suggested that it may be this sort of cooperative phenomenon that is the long-sought resolution of the binding problem. Whether or not this theory turns out to be the ultimate solution to the binding problem, it's clear that scientists are on the road to unraveling the functioning of the brain, neuron by neuron.

The Neurological Program

Enough research has been done so that we can get a glimpse of what the future holds for our understanding of the brain. At the large-scale level, a millimeter-by-millimeter map of the functioning cortex will surely be done—in fact, I would be astonished if it took more than a decade to complete the job. At the end we will be able to take any mental activity—seeing the color blue, thinking about grandma, doing long division—and say precisely which areas of the brain light up while it is going on.

Roughly speaking, there are 100,000 neurons beneath each square millimeter of the brain's surface. This means that the deeper level of mapping—the one that involves individual neurons—is a much more daunting task. This situation, coupled with the relatively sparse amount of information we have at present, means that completing a neuron-by-neuron map of mental activities is probably going to take a generation or more to complete.

Nevertheless, as the example of visual processing shows, it is possible in principle to work out what each neuron in the brain is doing when any specific mental activity is going on. Let me call the construction of such a neuron-by-neuron map of brain activity the "neurological program." Its goal is to do for every possible mental activity what scientists have already done for the preliminary steps of visual processing.

There are many barriers to the completion of the neurological program, and the extreme complexity and interconnectedness of the brain is only one of them. I suspect, for example, that financial constraints may play a much more important role in limiting our knowledge of the brain than most scientists realize. I come from a background in high-energy physics, for example, a field that has a dream as ambitious as the neurological program. In this field, a single congressional vote ended the so-called Superconducting Supercollider project, in effect shutting down a research effort whose roots can be traced back to the ancient Greeks. Because of this experience, I don't have high hopes that funding can be maintained for the neurological program for the decades it will take to complete.

Having said this, however, I will argue that for our purposes the question of whether the program *will* be completed is a good deal less important than the fact that it *can* be completed. In what follows, I will speak of the neurological program as an established fact and assume that it is indeed possible to give a specific description of what neurons are firing when any mental activity is going on. As we shall see, if this statement turns out to be wrong (as it well might), my final conclusions will only be strengthened.

7

How Did We Get to Be So Smart?

The Evolution of Intelligence

It seems these two hunters were walking through the woods when they encountered a very angry (and very hungry) grizzly bear. One hunter started dropping his gear on the ground.

"What are you going to do?" the other asked.

"I'm going to run."

"Don't be silly—you can't run faster than that bear!"

"I don't have to run faster than the bear. I just have to run faster than you."

—ANONYMOUS

If we have learned anything in the last two chapters, it is that the brain is an organ of almost unbelievable complexity. The question we have to ask, therefore, is how such a system as the brain could have arisen in the course of evolution.

To understand why this question is so perplexing, you have to realize that the evolutionary game is played by a very special set of rules. Looking at human beings today, it is obvious that possession of a highly developed cortex has survival value for our species. It allows us to build tools, to develop language, to modify our environment, and to deal with all manner of changes in that environment. In the evolutionary game, however, it is not enough to say that having a finished brain would be a good thing. In order to answer the question I am posing, you have to be able to show how such a brain could have developed over a period of time. After all, there is no way an individual *Australopithecus* could possibly have known that three million years after his death a creature with a much larger cortex would dominate living things on the planet. *Australopithecus* was concerned only with individual survival—with running faster than the other guy.

The Rules of the Evolutionary Game

It was the peculiar genius of Charles Darwin to see, in the bewildering variety of living forms on the planet, the operation of a single grand principle—the principle of evolution by natural selection. The story of the bear and the hunters is actually a good illustration of that principle. To see why, imagine moving forward in time thirty or forty years from that meeting in the forest. The hunter who could run faster would have survived and would be surrounded by children and grandchildren who would carry his genes, including whatever genes allowed him to survive the encounter with the bear. The slower hunter, alas, would not have left any descendants. Over time, if these genes continued to confer a survival advantage, they would spread throughout the entire population. This procedure, which is called *natural selection*, is responsible for the long-range development of living things on our planet. We can see this development in the fossil record, starting with ordinary bacterial pond scum 3.5 billion years ago and coming right up to the present.

The important point about natural selection, however, is that it operates on *individuals*. It does not operate on groups (except through operating on individuals), nor does it operate on individual genes.* Furthermore, it involves no moral judgment of any kind. Indeed, the slower of the two hunters might well have been an admirable person. He may have given money to charity and helped little old ladies across the street, while the fast runner could have been a real s.o.b. Natural selection doesn't care. Natural selection simply asks which of these two individuals will survive to have children. The genes of whichever one survives are passed to the next generation. Period.

When you're talking about something like running, it's not hard to imagine environments in which being able to run faster would confer a definite survival advantage. Animals who are able to run faster are more likely to catch their prey if they are predators, or to escape from their predators if they are prey. Consequently, in the language of evolutionary theorists, we say that there are strong evolutionary pressures to make members of a given species run faster in those environments.

If the situation changes, however, the pressures of selection change as well. For example, as soon as a major portion of the population can run faster than its predators, a point of diminishing returns sets in. There is no advantage to being able to run faster than the other guy if you can both run faster than the bear. In this case, the drop in selective pressure comes from the evolutionary process itself.

---

*I should warn you that there is controversy in the scientific community on this point. My statement represents the classical theory of natural selection, but there are those who will argue for both group and gene selection.

More often, the physical environment itself changes. For example, an insect having coloring that blends with a certain type of tree may escape the eye of predatory birds. In this case, natural selection will favor that particular coloring pattern. If, however, a blight comes along and wipes out that particular kind of tree, the advantage disappears. In fact, when those insects appear on the branches of other trees they may stand out, so that what used to be an advantage becomes a disadvantage. In other words, specific physical traits are not good or bad in and of themselves, but become good or bad in relation to the environment in which the organism finds itself.

The rules of the evolutionary game are simple. In order to be passed on to the next generation, a particular trait must give an advantage to a particular organism in a specific environment. If that condition is met, that particular trait will be selected for as long as the environment does not change.

All of which brings us back to the question of how the brain developed. As is the case with many other organs, it is easy to see that the finished product confers an advantage. But as we now know, that's not enough. Our brains are the products of millions of years of evolution. Millions of our ancestors had brains that were less complex and less accomplished than our own. In order for our brain to appear, every single change that was necessary to get to the present from the primitive brain of *Australopithecus* had to confer an advantage on the individual who first possessed it. Miss one link in that chain, and the entire structure falls down.

This is, of course, a general feature of evolution by natural selection. However, there is one caveat about this process of chain building. As we saw in chapter 2, the characteristics of any organism are coded in the DNA molecule. Changes in DNA will change the characteristics of the organism, and this in turn will affect that organism's ability to survive and reproduce. What matters is that the changes that follow from a given mutation provide, on balance, an evolutionary advantage. Thus, some accidental changes may survive because they are genetically linked to other traits.

Let me give you an example from another area of evolution to show how an evolutionary chain could be constructed. The ability to fly has obvious survival advantages, if only because it opens up new ways to gather food and avoid predators for the organism that can do it. So advantageous is the ability to fly that it has arisen independently many times in the course of evolution. Insects and birds, for example, fly in quite different ways because each represents a different evolutionary "discovery" of flight.

While it is possible to see how a fully developed wing could confer an advantage, it is hard to see how half of a wing (or a third or a tenth) could do so. Yet to make the chain from the original ground-based organism to an organism that flies, you have to supply each of those missing links. How can you do it?

There is actually a very interesting theory about the evolution of flight in insects. The idea is this: The original "wing" on an insect would have been little more than a bump located on the sides of the insect's body. That bump would not have allowed it to fly, or even to glide. However, it might have helped with other important functions. For example, cold-blooded creatures like insects have to exchange heat with the environment all the time. The suggestion is that the original bumps played the role of cooling fins—they increased the surface area of the insect's body and allowed it to radiate and absorb heat more efficiently. In an environment where it was important to get rid of heat (a desert, for example) or absorb it more efficiently (such as a colder climate) it is not hard to see that having bumps along the side of the body might confer an evolutionary advantage. Furthermore, it is not hard to see that the larger the bump, the more advantage you get. Thus, in the language of the evolutionary theorist, there would have been an evolutionary pressure for increasing the size of the bump along the side of the insect's body.

Eventually, of course, the clumsiness of moving the fins around would probably have canceled any advantage from having them be larger. It turns out, however, that at about that point the fins are large enough to allow the insect to glide. Suddenly, there is an entirely new ecological niche available to the insect. Instead of crawling around on a single tree, it can now glide from tree to tree in search of food and to flee from predators. Consequently, what was originally a cooling fin now plays an entirely different function—one that allows the insect to perform rudimentary flight. Once that threshold is crossed, the development of full wings is not hard to imagine.*

This process, in which a particular organ is useful first for one purpose, then another, occurs repeatedly in the history of evolution. I call it the "evolutionary switcheroo." We'll encounter it many times in this discussion.

The process of forging a link from ancestor to finished product, then, need not involve a continuous improvement in a single function. At each point in time, an individual organism faces the problem of survival—the hunters meet the bear. All that matters is that the individual has certain characteristics on which natural selection will act. Whatever effect natural selection is going to have will be made on the materials at hand, on the individual organism as it exists at the time. Evolution works on whatever is available and modifies it into whatever gives the individual who possesses it a survival advantage. This is what is meant by "survival of the fittest."

It is this aspect of evolution, in fact, that gives rise to many of the anomalies we see among living things. Perhaps the best known of these is the panda's "thumb," as pointed out by Stephen Jay Gould in his book *The Panda's Thumb* (W. W. Norton, 1982). The ancestors of the panda (which is distantly related to the raccoon) walked on all fours and, like dogs and cats, eventually lost the original thumb. When the environment in which the panda's ancestors found themselves changed to a bamboo forest, the panda had need of a thumb to strip leaves off the bamboo.

---

*An alternative scenario has the fins initially helping the insect skim across the surface of water, but the final outcome is the same.

What happened was that a slight bump on the wrist bone began to be enlarged. Even a small lump would let the animal strip the bamboo more efficiently and therefore exploit the energy resources of its environment better. Eventually, a spur grew out of the panda's wrist to play the role of the lost thumb. Obviously, this is not the bamboo-stripping system you would design from scratch, but it is a system that is very much in keeping with the spirit of evolution by natural selection. Each individual in the chain, from that original raccoonlike organism to the modern panda, received an evolutionary advantage from having a slightly larger growth of that bone.

The design of the human eye presents another such anomaly. You may recall that ganglion cells carry out the initial processing of the visual signal. The astonishing thing is that these cells actually sit in front of the cells that receive the incoming light—in effect, they cast shadows on the light receptors. No engineer would design a camera so that the machinery of the camera sat in front of the film or the photoreceptor. Thus, the human eye is also a very good example of evolution by natural selection. (In passing, I should point out that the construction of the eye with the ganglion cells in the front of the retina is not an evolutionary necessity. The octopus, which as we saw in chapter 3 is a highly visual organism, has its eye designed "right"—that is, the cells that process its visual input are located in back of the retina, rather than in front of it.)

The point here is that just as the evolutionary process is under no obligation to be moral, it is under no obligation to be perfectly efficient. Evolution produces organisms that are good enough to survive—not necessarily the organism that good engineers would build if they were starting from scratch. You never have to run faster than the bear to get your genes into the next generation, you only have to run faster than the other hunter.

As is often the case in evolutionary theory, we don't know enough about the environment in which our remote ancestors lived to be able to give a clear explanation of why the ganglion cells are placed where they are. There may have been something in the early environment that made it advantageous for them to be placed that way. On the other hand, as we discussed above, they might have been placed that way in a collateral accident associated with the development of some other feature that conferred a survival advantage. For example, the same genetic change that put the first primitive ganglion cell in front of the first primitive retina may have allowed the development of a more efficient lens. Someday, I suppose, all of these stories will be sorted out, but for the moment, we simply note that when they are, they will have to present us with the kind of unbroken link in the chain that we described above.

To understand how a complex organ like the brain could have developed, we have to show that at each step in the evolution from the distant ancestor to the modern organism, each change in the DNA conferred an evolutionary advantage on the organism in the environment in which it found itself at that time. Nothing less will do.

The Evolution of Intelligence

So now we come back to our original question: How did the human brain develop in a world governed by the rules of natural selection? How do you get from *Australopithecus* to someone who can write a symphony or prove a mathematical theorem by a series of steps, each of which confers a clear survival advantage?

Several crucial difficulties confront scientists who try to shed some light on this question. For one thing, as we saw in chapter 2, hominid fossils are rather light on the ground—there just isn't a whole lot of fossil data on early hominids available to us.

More important, however, the kinds of things we'd have to look for to answer this question are very difficult to discern from fossils. As we have seen, the brain functions like an interconnected set of villages, with different mental functions highly localized. A fossil preserves only the shape of a particular skull, including bumps and indentations on the inner surface that may give some insight into the gross structure of the brain that once occupied that skull. A fossil skull cannot, however, give any information about how the neurons in that brain are connected, or about the existence of particular groups of connected neurons performing specialized functions deep inside the brain.

Of course, as we indicated in chapter 6, some general aspects about brain function can be inferred from the shape of the skull. The high forehead of modern *Homo sapiens*, for example, results from the massive expansion of the frontal lobes—the seat of higher mental functions. The bulge in the back of the skulls of many primates (humans included) covers the occipital lobe, where visual processing is done. Thus, it would be reasonable to suppose that an animal possessing this bulge would have a highly developed visual system.

But we're really not able to go beyond these sorts of generalities on the evidence of the fossils themselves. The story of the evolution of human intelligence, then, is rather more speculative than most parts of evolutionary theory. It rests on indirect evidence and, not to put too fine a point on it, liberal use of the educated guess. For what it's worth, here are a few of the notions currently in vogue.

There is a consensus that the move to upright posture played an important role in the evolution of the human brain. Once the hands are free, adaptations such as grasping, throwing, and toolmaking become possible and capabilities on which natural selection could act. But why walk upright in the first place?

Richard Leakey and Roger Lewin, in their book *Origins Reconsidered*, suggest one way upright posture might have arisen. Thirty million years ago, most of Africa was covered by rain forest, home to at least twenty different species of great ape. For reference, today the Earth holds four such groups—chimpanzees, gorillas, orangutans, and humans.* About that time, however, tectonic processes deep in the Earth began to pull the continents apart. This process is still going on, and the Red Sea and the Great Rift Valley are modern results of it.

As a result of the motion of the tectonic plates, the climate in Africa began to change. The forests began to disappear, being replaced first by stands of trees separated by open plains and finally, today, by savanna. When the area was in the intermediate state of separated forests, the ability to move from one forest "island" to another would clearly have survival value—think of what would happen if food gave out in one island, or predators suddenly appeared.

It is likely that at the time at least some species of great apes had developed the ability to walk short distances in an upright posture. The advantage of being able to move quickly over the ground for short distances (between trees, for example) is obvious. We know, for example, that modern chimpanzees can do this, effecting a rolling gait with their arms held above their heads for balance. Given a population of apes with that ability, and a changing environment, it's not hard to see that natural selection would have acted to drive upright posture.

In this example, the move to upright posture illustrates many of the points we made about the rules of the evolutionary game. First, there was a major change in the physical environment, followed by the extinction of many of the existing species. The species that survived did so by modifying existing structures to cope with the new situation. The result: the only great apes who walk upright.

But, as was the case in the conversion from cooling fins to wings, once this change occurred, new possibilities for natural

---

*As usual, there is debate in the scientific community about the details of this sort of grouping. Some scientists, for example, would include the gibbon in this list. For our purposes, groupings don't matter—there are a lot fewer now than there were then.

selection opened up. The stage was set for another evolutionary switcheroo. Scientists argue that the evolution of human intelligence, like the evolution of insect flight, may well provide an example of a seemingly unintended benefit in one area from developments in another.

William Calvin, a neurophysiologist at the University of Washington, has proposed an intriguing scenario for how the switcheroo might have worked. It's based on the assumption that there is a region of the brain, presumably in the left hemisphere near the language centers, involved in dealing with planning and analyzing sequences—the sort of sequences involved in stringing together words into sentences. The initial development of this ability, argues Calvin, came from the obvious advantage accruing to individuals who are able to throw things accurately.

The ability to throw a rock (to take one example) involves something called *ballistic movement*—a rapid movement of the arm and hand. It turns out that if a motion takes less than a fifth of a second to execute, there isn't time for the brain to make corrections while the motion is going on. All the movements have to be planned in advance, then executed. An individual who was able to figure out the movement involved in throwing would be more likely to obtain food and thus survive to make sure the genes involved are in the next generation.

Later on, this ability to plan movements would have been involved in the production of tools. Flint knapping—the production of stone tools—requires the same sort of movement of the arm as throwing. In fact, people who are good at it, such as my father-in-law Vern Waples, who practices the art as a hobby, say that you actually "throw" the rock in your hand at the rock being chipped to make cutting tools and arrowheads. My own experience as a carpenter leads me to conclude that the same skill is used in hammering nails—a good carpenter "throws" the hammer at the nail head.

So there were plenty of pressures in the environment of early hominids to drive the ability to plan ballistic movements. Then,

Calvin argues, another evolutionary switcheroo took place. The planning ability developed for this purpose was coopted to help humans develop language (which involves stringing sounds together into words and words together into phrases and sentences) and other higher mental functions. Of course, before we can accept this notion, we have to have a fully worked out step-by-step account of how the switcheroo took place.

Nevertheless, I have to admit that I have high hopes for this theory, if only because it explains something that, to me, is one of the great mysteries of evolution—human musical ability. No matter how hard I try, I can't think of a single evolutionary pressure that would drive the ability of humans to produce and enjoy music and dance. As a long-time performer and student of European folk dance and a long-time opera buff, this has always seemed like a serious problem to me—perhaps a more serious problem than that perceived by most of my colleagues. In Calvin's theory, however, music and dance—the ability to string notes and movements together into a harmonious whole—emerges as a consequence of the ability of some *Australopithecine* to knock down a fast-moving rabbit with a rock. Most satisfying.

Could Human Intellectual Capacity Be Unique in the Animal World?

Homo sapiens evolved from some primordial great ape by a process that follows the same rules as any other evolution. How then could humans be so different from everything else?

This question and others like it illustrate a common misconception about the way the universe works. The assumption is that processes following the same law ought to produce the same results. Nothing could be further from the truth. Consider, for example, the fall of two meteorites onto Earth. Both trajectories are governed and predicted by ordinary Newtonian mechanics. Yet one may fall in the ocean, the other on your house. Same laws, different outcomes.

In just the same way, the process of natural selection, acting over millions of years, can produce many unique products. The use of sonar by bats, the infrared sensors of pit vipers, and the trunk of the elephant are all examples of one-of-a-kind products of natural selection. Why shouldn't human intelligence be added to this list?

In fact, Steven Pinker, in his book *The Language Instinct*, pokes fun at the notion that evolution cannot possibly produce unique structures like the elephant's trunk or the human cerebral cortex. It turns out that the trunk of the elephant is a truly remarkable organ, containing no fewer than sixty thousand distinct muscles and capable of an immense range of motion, from lifting logs to writing on a blackboard with a piece of chalk. Like humans, the elephants have no living relatives that resemble them—the nearest is an animal called the hyrax, which has something of the appearance of a guinea pig. Pinker asks us to imagine what elephantine scientists would do if they were bent on showing that their species was really no different from their nearest neighbors:

> They would first point out that the elephant and the hyrax share about 90 percent of their DNA and thus could not be all that different. . . . Though their attempts to train hyraxes to pick up objects with their nostrils have failed, some might trumpet their success at training the hyraxes to push toothpicks around with their tongues, noting that stacking tree trunks or writing on blackboards differ from it only in degree.

In the end, there is no reason why the human cerebral cortex cannot take its place alongside other unique organs in the animal kingdom. In terms of the dilemma set forth in the first chapter, this means that we should have no qualms about assigning a special place to our species based on the evolutionary development of the cerebral cortex.

But, as we pointed out, this conclusion forces us to encounter the second horn of the dilemma—the possibility that computers designed through the use of that same cerebral cortex may someday duplicate and even replace it. It is to that subject we now turn.

8

Moving Wheels and Moving Electrons
How a Computer Works

I think we discovered something today.

—Physicist John Bardeen to his wife on the
day the first transistor was built

Next time you're in your car, I'd like you to pay some attention to the little display of numbers on your speedometer—the one that tells you how far the car has been driven. You will notice that the display consists of a set of numbers that record tenths of a mile, miles, tens of miles, hundreds of miles, and so on, starting from the right as you look at the numbers. The device (called an odometer) works like this: There is a cable attached to the transmission of your car that spins around as the car moves—the faster you go, the faster it turns. The cable is attached to a gear at the

113

right of the odometer in such a way that every time the car travels a tenth of a mile, the gear makes a tenth of a turn. The tenth-of-a-mile display consists of a series of numbers painted on this gear, and you can watch the gear turning as new numbers click into place. When the tenth-of-a-mile gear has made a complete turn, you have traveled a mile. The odometer gears are arranged so that when this happens, a connection is made to the mile gear, which then ratchets up one tenth of a turn. When the mile gear has made a full turn, the ten-mile gear makes a tenth of a turn, and so on. As you drive, you see a steadily progressing display of numbers on the odometer.

This is a machine that takes input (the turning cable), manipulates it (by means of the gears), and presents its results as an output (the numerical display on the odometer). By mechanical means, it performs a specific arithmetic operation (addition), and by the connections between the gears it does the arithmetical operation we call *carrying*. It expresses a number (how far you've traveled) in terms of physical quantities (the positions of the gears). It is, in fact, the descendant of over three hundred years of automatic calculating machines, the forerunners of the modern computer. The gears are set when the device is built, and they keep performing the same numerical addition until the car is junked— by their nature, they can't do anything else.

If you think about the odometer as a paradigm of a calculating machine, however, you realize that there's no particular reason why either the input or the manipulations have to involve mechanical contrivances like gears and cables. You could equally well represent numbers by pulses of electrical current and manipulate them by electrical means. In this case, you would be calculating by moving electrons rather than by moving wheels. This, of course, is how modern computers and calculators work, but to get from the moving gears to moving electrons, we have to talk a little about how we can represent numbers by electrical impulses.

You may be surprised to learn that the numerical scheme used for representing numbers in modern computers is quite old—it was invented by Gottfried Leibniz, the co-inventor of the calculus. It is called *binary arithmetic,* and you can understand it by thinking about our ordinary decimal number system. Normally, we start counting by running through the numbers from one to nine, then go to the next number by writing ten—by putting a one in the tens place and starting the counting process again. The reason we use this particular system is undoubtedly related to the fact that we have ten fingers, but it's not the only possible system. The ancient Babylonians, for example, used a system based on counting to sixty (in their system, the number eleven would be sixty-one in ours). The fact that we still divide a circle into 360 degrees is a relic of the old Babylonian number system.

Binary numbers have only two digits—a zero or a one. Instead of counting up to nine before we start again, in the binary system we count only two numbers (zero and one) and then move to the next place. In the binary scheme, the number one is *1,* the number two is *10,* the number three is *11,* four *100,* and so on. As we shall see in a moment, this makes binary numbers ideal for use in modern computers.

As a historical aside, I should note that Leibniz, who was actually interested in the problem of building calculating machines, never thought of putting together his binary numbers with the machines. Some historians have speculated that had he done so, we might have seen gigantic steam-powered mechanical computers as a part of the nineteenth-century industrial revolution. If any of my readers are science fiction writers, I heartily recommend this as fertile ground for a new novel. As it turned out, however, Leibniz's only use of binary numbers was to produce a weird "proof" of the existence of God. (It had something to do with representing God by the *1* and a universe without Him as a *0.*)

When numbers are represented by a continuous quantity like the angle through which a cable or gear has turned, the machine is referred to as *analogue.* If the numbers are represented as digits or as ones and zeros, we say the machine is digital. Although there are analogue computers around, by far the greatest majority are of the digital type. And while digital computers can be (and have been) built using many different kinds of working parts, almost every computer you're likely to encounter will have things called *transistors* as their basic working unit.

Just as we began to understand the brain by talking about the neuron, we will start our description of the computer by talking about the transistor.

The Smallest On-Off Switch

The transistor is a device that was invented two days before Christmas 1947 by John Bardeen, Walter Brattain, and William Shockley. It was designed to replace a device that only old-timers will recognize these days—something called a vacuum tube. (Remember, though, that mechanical calculating machines existed before either the vacuum tube or the transistor was dreamed of.)

Transistors are made from materials called *semiconductors*, the most familiar of which is *silicon*—one of the elements that makes up ordinary beach sand and window glass. The silicon atom has four electrons in its outermost orbit. Think of these outermost electrons as hooks that can connect one atom of silicon to another. In a pure silicon crystal, each of the four hooks on each silicon atom is linked to one of the hooks on another silicon atom, and the whole thing forms a strong solid crystal. In theory, a material like silicon should not conduct electricity, since the electrons are hooked together and are not free to move. It happens, however, that the normal vibrating of atoms in the crystal is enough to shake a few electrons loose. Thus, the silicon has a few electrons that are free to move, and these electrons can constitute an electric current. There are not, however, nearly as many free electrons in silicon as you would find in a metal like copper, so there isn't much of a current. This is why it's called a semiconductor—silicon conducts electricity all right, but not very well.

By a process called *doping*, small amounts of other elements can be mixed in with molten silicon to produce semiconductors with different properties. In essence, it is possible to produce semiconductors in which some of the impurities, once they are locked into the structure, have a positive electrical charge and other types of semiconductors in which other impurities have a negative charge. Thus, there are two types of doped semiconductor, called *p* and *n*, respectively, depending on which type of impurity was added into the melt when the silicon solidified.

The simplest transistors are a "sandwich" of semiconductors. If the "meat" of the sandwich is a *p* type semiconductor, then the two slices of "bread" are *n* type, and vice versa. This simple structure allows one to have an enormous control over the amount of electrical current that flows through the device. In computers, the transistor is used as a switch—things are arranged so that electricity flows through the sandwich (*on*) or in such a way so that the flow of electrical current is blocked (*off*). The basic technique for doing this is to run electrons onto the "meat" of the transistor until the amount of negative charge is high enough to stop the flow of electrons through the device. In this case, no electrical current can flow and the transistor will be *off*. Similarly, if the electrons are removed from the "meat," electrical current will be able to flow and the transistor will be *on*.

The best way to think of a transistor when it is used in this mode is as something that operates the same way as a valve in a water line. You can have a lot of water flowing through a pipe, but a small amount of energy applied to the handle of a valve can turn that water on or off (you do this every time you use the faucet on your sink). Turn one way and you open the valve and allow the water to flow. Turn it another way, and you close the valve and stop the flow. The water either flows or it doesn't. In just the same way, current either flows through a transistor in a computer or it doesn't.

For completeness, I should point out that there are other ways that transistors can be used (they are the main workhorses of the amplifiers found in any radio or TV set, for example). Furthermore, the "sandwich" transistor that I've described is actually one of the first types of transistors that was ever built. Today, there are many other designs for transistors. Nevertheless, the main point—that a transistor can be turned on or off by manipulation of small numbers of electrons—remains true for all of them.

Transistors, Information, and the Digital Computer

The main reason that transistors can be put together into a machine like the computer has to do with the nature of information. All information, whether it concerns written words, musical notes, or the future state of the Earth's climate, can be represented by *bits* of information. A bit of information is the answer to a simple question—yes or no, up or down, on or off. We call this type of information *digital*. Since the transistor is a device that can be operated in such a way that it is either on or off, you can see that by its very nature it is suited to dealing with digital information. If you think about it for a moment, you'll realize that the natural way to represent digital information is through the use of binary numbers—there is a natural affinity between on or off and one or zero. Digital information thus appears as a string of zeros and ones. If you think of each zero in the string as a transistor switched off and each one as a transistor switched on, you can see that there is a clear correspondence between information and arrays of transistors.

Let me give you a simple example to show how bits can be used to convey information. Suppose you want to give someone a rough indication of the temperature in a particular city. Suppose further that you knew that the temperature would be between 40 and 80 degrees and you only wanted to be in the right decade—that is, you want to tell the person that the temperature is in the thirties, but not differentiate between 36 and 37 degrees. If you had two transistors, there would be four possible ways that those transistors could be arranged: They could both be on, they could both be off, the first could be on and the second could be off, and the first could be off and the second on. You could then make a code that would say something like this: If both transistors are on the temperature is in the seventies, if the first is on and the second is off the temperature is in the sixties, if the first one is off and the second one on then the temperature is in the fifties, and if they are both off the temperature is in the forties. By specifying two numbers—a zero or a one for each transistor—you could convey the information about temperature. Although it is not obvious, a more complicated string of numbers could convey anything from a television picture to a phone conversation.*

Therefore, the working part of a computer can be thought of as a system of transistors that can be turned on or off at will. Different configurations of transistors correspond to different information content, and the ability to turn the transistors on or off corresponds to the ability to manipulate information.

A machine like this differs fundamentally from the odometer with which we started this chapter because it doesn't have to do just one thing. By adjusting various voltages on individual transistors, for example, it is possible to change the way they work. Run a certain number of electrons onto the "meat" of a transistor with one voltage setting and you might turn the current off. Run the same number of electrons on at another voltage setting, on the other hand, and the current might keep running. In everyday language, we say that it is possible to *program* the computer—to give it instructions that change the way it manipulates information. It is this flexibility that makes computers so important in our technology today.

In the machine I am using at the moment, for example, my keystrokes send electrical signals to the computer (input data),

---

*A more complete description of how various types of information can be expressed in terms of bits is given in my book A *Scientist in the City* (Doubleday, 1992).

and the word processor in the machine (the program) manipulates them to produce the text. If one bit of information in those transistors were to change, the letter it represents in the code of the word processor would change as well. Thus, the word *cure* might change to the word *care*.

A word of warning: A real computer in the real world is much more than a set of transistors, just as the brain is much more than a set of neurons. What I have described above is what would normally be called the *central processing unit*, or CPU, of the computer. This is the place where information is actually changed around and manipulated. Computers will also have a place where information can be stored. This is called the *memory*, and comes in many different forms. In the memory, information is not stored in transistors but in magnetized material such as a tape or disk, in which little grains of iron compounds act as tiny magnets. What would correspond to a transistor being in the *on* position might be something like "north pole of the little magnet pointing up," and what might correspond to *off* might be "north pole of the little magnet pointing down." Information is retrieved from memory when it is needed, manipulated, and then returned to the memory. In principle, however, the distinction between CPU and memory need not concern us in what follows.

A more important distinction is the one between the actual physical structure of the computer (what is called the hardware) and the instructions that tell the machine what to do (the software). A set of instructions about how to solve a specific problem is called an *algorithm*.

The Turing Machine

In 1937, the British mathematician Alan Turing proved one of the most fundamental theorems in computer science. He showed that the operation of any calculating machine operating an algorithm, no matter how big, how complex, or how expensive, could always be represented by the operation of a simple machine—a machine that has since come to bear his name. The Turing machine allows us to think about computers in an abstract, general way, without reference to any specific machine. I should stress, however, that the Turing machine is strictly a hypothetical construct—no one has ever (or is ever likely to) build one.

A Turing machine consists of two parts. The first part can be thought of as a long tape marked off into little squares. Each square contains one bit of information—think of it as either a zero or a one. The second part of the machine is a mechanical device. You can either think of the device as something that moves along the tape or as something that sits still and has the tape fed through it. In any case, the mechanical device has instructions (a "program") that tell it what to do when it encounters each square on the tape. For example, when a particular square on the tape comes into the machine, the instructions may say, "If it is a zero, change it to a one, if it is a one, change it to a zero." The appropriate change will be made on the tape, which will then exit the Turing machine.

Now it is important to realize that even in theory the Turing machine is not the same as a real computer. As I am typing these words into my word processor, for example, what is happening is that each letter is being recorded in an array of eight transistors (eight bits of information are called a *byte*). Periodically, the information in those transistors is transferred to magnetic storage, either in a hard disk or a floppy. The operation of any real computer can be described in a similar way, from the largest supercomputer to the littlest microchip in a kitchen appliance. There doesn't seem to me much connection between this machine and some tape running through a box.

The point, however, is that Turing showed that the net effect of the operation of any complex real computer can be represented by the operation of one of the machines that now bears his name. Thus, if your main concern is understanding the abilities and limitations of computers, you need to examine only the Turing machine to find them. Once you have done that you can be assured that the abilities and limitations of any real machine will be the same.

Neural Nets

Computers, in the end, are simply collections of electrical devices. Many commentators argue that this implies that computers are just sophisticated versions of things like typewriters or adding machines (I regret that I used to be included in this number). To be honest, this statement describes most operating modes of most computers. A set of instructions, called a *code* or a program, and some input data are provided for the machine, and the machine manipulates the input data according to the code.

I could, in principle, take detailed information about the design of the keyboard of the machine on which I am writing, the CPU of the computer, and the word-processing program and predict exactly what the computer will do in any situation. If I misspell a wird, it's no good blaming it on the computer—it's just following my directions. In this sense, this particular computer is being used in a mode not much different from a typewriter.

But over the last few decades, computer scientists, chafing at the restrictions imposed by typewriter-style operation, have started to put together computer programs that are self-consciously intended to mimic the operation of a real nervous system. Bearing names such as "neural nets," or "learning machines," these computer systems can produce surprising—even startling—results. The simple game I described in the Preface—the one that found the "reeool" for choosing figures—was played on a machine using this approach to operation. Neural nets have the unique property of allowing computers to do something very much like what humans and other animals seem to do when they learn from experience.

As often happens when we talk about things like learning, we have to go back to some fairly primitive animals to see how it works. In this case, the animal is a sea slug, a shell-less genus of mollusk called *Aplysia*. About the size of a child's football and equipped with a relatively simple nervous system, *Aplysia* has become something of a workhorse for studies of animal behavior.* The most studied response is the slug's reflex withdrawal when it is touched in the region of the gills. When the animal learns this response, there is a selective strengthening and weakening of the synapses in its nervous system. By a process we don't fully understand, but which seems to involve increased neurotransmitter production and changes in both the pre- and post-synaptic neurons, the synapses that are involved in causing the animal to pull back seem to become easier to excite with each trial. Thus, the *Aplysia* nervous system seems to adjust its own operation as a result of experience. We believe that learning in humans, though undoubtedly more complex, operates the same way at the level of neurons.

Neural nets are an attempt to design a computer program that can operate in the same way. The essential point is that the weight given to different input data can be changed to respond to the success of the program in carrying out its goals, an analogy to the way the synapses can be strengthened or weakened in real neural systems. The goal is to build a system that can "learn" the way a real nervous system does.

I should point out that neural nets are not some sort of "blue sky" idea that exists only in the minds of theoreticians. They have actually been built and are being widely applied. They are used in air traffic control (for aircraft recognition), finance (in screening credit card transactions for possible fraud), security (fingerprint and voice recognition), and medicine (image processing and diagnosis), to name just a few applications.

---

*At the same late night beer fest where I heard about Kanzi (see page 59), I was told of one member of the animal behavior community who has developed a recipe for cooking his *Aplysia* after the experiments are done. His dish apparently tastes something like paella, a Spanish seafood dish.

In thinking about neural nets, however, it helps to have a specific example in mind; therefore, let me talk about the problem of recognizing a pattern in a visual field—reading handwritten zip codes on envelopes, for example. (This particular technology is undergoing rapid development for the U.S. Post Office.) The neural net would consist of three parts—an input unit (in this case a set of phototubes, each of which scans one little square on the envelope), an output unit (perhaps a rendition of the zip code in electronic form), and, in between, a so-called hidden unit that converts input to output.

In this case, you might well have several different levels of transistors in the hidden unit, each taking signals from transistors in a lower level, combining them, and sending them on up to transistors in a higher level. For example, systems of transistors in the first level of the hidden unit may receive signals from many different phototubes—they might sense a current if the phototubes detected a light spot on the envelope and no current if the phototube detected a dark spot. The groups of transistors would weigh these signals differently (for example, they might assign twice as much weight to phototubes reading the center of the visual field as to those reading the periphery). In the end, the system would add up all the weighted inputs and by sending electrons into the "meat" decide whether a given transistor in the next level up is on or off. Higher up in the hidden unit, you might have transistors that only turn on when appropriately weighted transistors in a lower level indicate a visual field is dark on one side and light on the other. This arrangement should remind you of what we know about visual processing in primates, as discussed in chapter 6.

In the end, the hidden units will send a signal to the output and you get an answer—"the zip code is 90210," for example. In many neural networks, this result is then compared to the known correct answer. If the weights assigned to various connections in the network are more or less random at first, chances are that what comes out of the hidden unit won't be much like the written input.

The program is designed so that the weights are changed at this point (you might now assign three times as much weight to sig-nals from the center of the field as to those from the periphery, for example). The network tries again with the new weights, changes, tries again, and so on until it gets things right. This trial-and-error process is called *training*.

Eventually (and it often takes a long time), the weights in the network will be adjusted so that it gives correct answers for a variety of test inputs. We say that the network has worked through its "training set." The network with adjusted weights is then put to work reading patterns without supervision or teaching. (As of the publication of this book, for example, experimental neural nets operated by the Post Office are capable of reading over two-thirds of the zip codes on letters—even those that are handwrit-ten.)

In my days as a particle physicist, I saw the very beginnings of computer pattern recognition. In the 1960s and 1970s, the main tool for investigating the collisions of high-energy particles with targets was a device called a *bubble chamber*. The end result of an experimental run was a long reel of photographic film, with each frame showing the tracks of particles coming from a collision. These rolls were run through specialized projectors that showed the patterns on large, tabletop screens, and groups of people (called *scanners*) would watch the film go by and pick out events whose patterns fit those laid out in advance by the physicists. I can re-member large darkened rooms full of those tables and very bored scanners.

As you might guess, there were problems with this procedure— I know one physicist who routinely reran films that had been scanned on Monday mornings and Friday afternoons, on the grounds that scanners anticipating or recovering from weekends were not as attentive as they could be. It was experimental physi-cists who took some of the first steps in devising ways of automat-ing the scanning process, for the simple reason that they needed to replace the human scanners. It's a long way from those early attempts at pattern recognition to reading zip codes, but the story illustrates an important point: Tomorrow's cutting edge technology often arises in unpredictable ways from today's basic research.

For our purposes, however, the development of neural nets illustrates something very important about computers. It is possible for machines to do things for which they are not specifically programmed. No one actually gives a neural network explicit instructions about how to read a zip code. Instead, the network is told how to go through the training process and get to that point more or less on its own.

Moore's Law

If I had to point to one astonishing thing about the development of transistors, it is that they have become almost unbelievably miniaturized since that pre-Christmas day in 1947. The first transistor was almost the size of a golf ball—it would have been hard to put even one of them in a modern hand calculator. Today, of course, it is not uncommon to have hundreds of thousands of transistors on a microchip not much bigger than a postage stamp. The old machine on which I am writing these words, for example, has a chip that probably includes a few hundred thousand transistors, but more modern machines might well have in excess of a million.

Just as an aside, have you ever thought about how astonishing it is to be able to own a million of any manufactured item? If you went out and bought a million paperclips retail, for example, it would probably cost you almost as much as a laptop computer.

The first computer with a microchip was made in 1971—it was the Intel 4004 and contained 2300 transistors. Today, it is routine for microchips to contain millions of transistors. The Intel P6, for example, has some 5.5 million, and some projections call for 100 million transistors on a chip by the year 2000. It was Gordon Moore, one of the founders of Intel, who first noted that all of the indices of quality on a computer—the number of transistors on the size of a chip, the size of the memory, and so on—seemed to double every two years. Called "Moore's Law," this seems to be a pretty good rule of thumb* for computer development.

Moore's Law seems to hold regardless of the changes in technology needed to produce each new advance. It has survived the changes from mainframes to minicomputers to PCs, and I would be surprised if it didn't continue into the future.

Extrapolating from the past, you have to conclude that the day when we will be able to put 100 billion transistors on a chip will come sometime around the year 2020. Thus, it is reasonable to assume that sometime in the foreseeable future the number of transistors that can be put on a microchip will approach or surpass the number of neurons in the human brain. This is something we will have to keep in mind when we compare these two systems.

---

*Incidentally, according to the *Oxford English Dictionary* there is no basis whatever for the story, beloved of feminist scholars, that this phrase originated in an English law that stated that a man could not beat his wife with a stick bigger across than his thumb. My own guess is that the phrase originated in rough carpentry, and is related to the fact that a man's thumb is about an inch across.

9

Artificial Intelligence, Learning Machines, and Chinese Rooms

If you don't think too good, then don't think too much.

—BASEBALL GREAT TED WILLIAMS

Artificial Intelligence (AI)

I guess I might as well get one thing off my chest in the very beginning of this chapter. One of the problems I have with people who claim that computers can perform all kinds of functions that we normally reserve for the human brain is their absolutely abysmal track record in the use of the English language. Time and time again, they will write a clever program that, to an indulgent

129

observer, might appear to do something that has a resemblance to a human mental attribute like intelligence, then turn around and talk about things such as "artificial intelligence," completely missing the point that what the computer is doing bears no relation at all to the functioning of the brain.

Francis Crick, for example, remarks that one of his major contributions to research on neural nets was to get the computer jocks to stop calling their collections of transistors "neurons." I can only wish that there were more people performing this sort of function in that field.

When I was a student, there was a standing joke that the only ability you had to have to go into the field of artificial intelligence was the ability to spell at least one of the words. Like most student jokes, this was a bit of an overstatement, but it contained a kernel of truth. Artificial intelligence is a field that has suffered for decades from hype and overselling.*

Now don't get me wrong. It is possible to make computers perform all manner of interesting and useful functions. The machine I described in the Preface, with which I played the rule-finding game, was an example of this sort of thing. It is even possible to make machines that can carry on an interesting dialogue or play chess at the championship level. None of these accomplishments, however, mean that the machine has "intelligence," at least in the sense that the word is normally used.

Let me give you an example both of what machines can do compared with the misuse of English I described above. One human mental function that is extremely hard to duplicate on a machine (or, for that matter, to understand) is the intuitive leap—the sudden inspiration that allows you to "get it." There are many problems whose solution depends on such an inspiration. The "brain teasers" that you encounter in Sunday supplements in your newspaper, for example, require exactly this sort of insight.

---

*To be fair, I should tell a joke from the other side, which is that "artificial intelligence is anything computers couldn't do five years ago."

A few years ago, I was told about an attempt to write a computer program that would display intuition. It's such a perfect example of the phenomenon I'm talking about that I'd like to tell you about it in some detail. The problem the experimenters chose to attack is called the "mutilated chessboard problem." The idea is that you take a chessboard, which has 64 alternate black and white squares, and remove two squares in opposite corners. You now have a chessboard with 62 alternate black and white squares. You are then given a set of 31 dominoes, each with a black square and a white square. The problem: Can you cover the entire chessboard with those dominoes, always laying black on black and white on white?

I have to admit that I hate this kind of problem. It's of such a trivial nature that I find it difficult to justify spending much time on it, particularly because I know that the answer depends on seeing a special trick. In this case, I'll spare you the frustration of trying to work out the problem by giving you a hint about how to get to the solution. Opposite squares of a chessboard are always the same color. This means that the mutilated chessboard will have thirty squares of one color and thirty-two of another. The thirty-one dominoes, on the other hand, have only thirty-one white squares and thirty-one black squares, so it is obviously impossible to do the covering problem.

When human beings attack this problem, they will typically go through a period of trial and error—laying the dominoes down first one way and then another. Eventually, though, they "get it" and see how the solution works. A computer assigned to do this problem will also start the random "laying down" process, but left to itself will just keep on doing it. In the case I was told about, after the computer had been fiddling around for a while, the experimenters told it to count the numbers of squares of each color—basically they gave it the same hint that I gave you as soon as I stated the problem. The computer then went on to solve the problem. Arguing that intuition is just a form of recognition, the experimenters then went on to claim that their computer program had demonstrated intuition.

I beg to differ with this conclusion. What it shows is that if you give the computer a certain fact, it can take that information into account. But the program did not come up with that fact by itself, which is what human problem solvers do. The program demonstrated that it could use the fruits of intuition but is about as far from real intuition as you can get.

This is not to say, however, that there is no use for artificial intelligence and machine learning. In fact, there are many areas where computers can be used to great advantage. Let me tell you about a couple of these that are often cited as examples of machines "taking over" human mental functions, and then tell you how the systems actually work. In doing so, I am playing the role of the stage magician explaining his tricks—you'll find that once you understand what's going on, the magic disappears.

Just Playing

Perhaps the most widely publicized accomplishment of AI is in the development of programs that can play chess. Chess is actually the perfect game for a computer to tackle. It has well-defined rules and a bounded space of possibilities but is still difficult enough to constitute a challenge for the best machines around.

It's easy to follow the progress of chess-playing machines because the governing bodies of the chess world maintain a rating system in which every serious player is assigned a number, and different numerical ranges correspond to different designations—expert, master, etc. The ratings are set up so that if player A's rating exceeds player B's by 200 points, player A will be expected to beat player B 75 percent of the time.

In 1975, computer chess programs had a rating of about 1500, roughly equivalent to the average rating of members of the United States Chess Federation. By 1985, they had achieved a rating of 2200, enough to earn the title of master. By 1990, their rankings were above 2400, so that they were playing at the human championship level. Then, in August 1995, the unthinkable happened. The human champion, Gary Kasparov, a man who many say is the best player the game has ever seen, had an off day and lost to a program called Genius 2. (It really was an off day—the program was later eliminated from the tournament by another human grand master.)

In 1996, in a challenge match with an IBM machine called Deep Blue, Kasparov was able to win, but not before he made everyone nervous by losing the first game.* Although humans are still on top for the moment, few people doubt that it is just a matter of time before the world chess champion is a computer.

The interesting thing here is not so much that a computer can or can't play chess better than a human, but how the computer plays. To understand computer chess, you have to think a bit about how a chess game can develop.

When a game starts, white can move any of eight pawns one or two spaces forward and either knight two ways—a total of 20 possible moves. Black can do the same. Therefore, there are $20 \times 20 = 400$ possible configurations the board can have after the first move. After that it gets a little harder to calculate possible moves because pieces like bishops and queens can move any number of spaces. For purposes of estimation, however, let's say each player has sixteen possible moves—one for each piece. Then at the end of the second move, there are $400 \times 16 \times 16 = 102,400$ possible configurations, over 20 million at the end of the third move, almost 2 trillion by the end of the fifth, and so on.

In principle, it should be possible to start with one possible move and examine each possible configuration that could result from that move many steps into the future. By going through such a calculation at any point in the game, you should be able to pick

---

*Deep Blue is an improved version of a machine called Deep Thought, whimsically named for the computer that answered the question about "life, the universe, and everything" in A Hitchhiker's Guide to the Galaxy by Douglas Adams.

the best move to make. Deep Blue, for example, can calculate over 200 million moves per second, which, with some technical "pruning," allows it to "see" seven moves into the future.

The point, of course, is that no human being plays chess this way. A grand master will typically assess a situation and follow only a couple of moves into the future, dealing with no more than 100 possible configurations in all. The brain seems to be doing something here beyond brute force computing.

The reason that chess computers have been getting better and better over time is that computational power—the ability to examine more and more possible variations of moves—has increased dramatically. The reason that I am confident that there will soon be a computer world chess champion is that computing power is going up at a sufficient rate that machines will soon be able to calculate dozens of moves into the future and pick the best possible strategy.

Does this make the chess-playing machine intelligent? Now that I've explained how the machine works, I suspect that most people would answer this question in the negative. What I would say is that if you want to use the word *intelligence* to describe this kind of ability in the machine, then you have to be very careful to understand that this is not the same kind of intelligence that we talk about when we are dealing with human beings. The machine may get to the same result, but it gets there by a different route. Furthermore, it does so in a situation in which the possible courses of action are extremely circumscribed and limited—a situation, in other words, that is very different from the kinds of things to which we normally refer when we use the adjective intelligent in real life.

Expert Systems

One use of artificial intelligence that is starting to have major economic and technological implications is the use of something called *expert systems* to deal with specific problems. As an example, let's consider a situation that arose recently in my home. An electrical circuit shorted out, tripping the circuit breaker and turning out the lights and wall sockets in several rooms.

The circuit breaker had tripped because too much electric current had been flowing in it, but what could have caused this? There were several possibilities, and I had to go through a logical procedure to decide which it was. Sometimes, for example, when a light bulb burns out there is a momentary surge that is capable of tripping the breakers. Alternatively, there might have been crossed wires somewhere in the circuit, or I might have had some appliance that had developed a short plugged into the circuit. I went through a series of tests to determine which of these had happened.

First I went down to the breaker box in the basement and turned the circuit breaker back on. If the problem had been caused by a light bulb in the process of burning out, the breaker would have stayed on. It didn't. Conclusion: That wasn't what was causing the problem. I then went around and unplugged all the lamps and appliances in the circuit. If one of them was causing the short, the circuit breaker would now stay on. It didn't. Conclusion: The short circuit was somewhere in the wiring. I then went around to the electrical boxes and began disconnecting parts of the circuit. (This is not something you should do unless you have had a good deal of experience working with electrical circuits. Better to call an electrician than to accidentally fry yourself.) I came to a box where the circuit breaker would stay on after a particular switch had been disconnected. Conclusion: The short in the circuit was somewhere in the branch controlled by that switch box. It happened to be a branch that turned off the lights outside the house. A quick inspection of the lights outside revealed that a recent major snowstorm had broken one of them and had flooded a region where there were wires, causing the short. With this knowledge I could then effect a repair, which in this case involved replacing the light fixture.

As shown in the diagram, this procedure can be thought of as a kind of tree, with specific questions at each branching. When you get to a node ("Does the circuit work when I turn it back on?"), there is a different path to take depending on the answer you get. This is typical of the way someone who knows how a particular system operates will go about dealing with a problem. A doctor diagnosing a disease or an automobile mechanic diagnosing a malfunction in a car will follow the same sort of logical tree (with different questions, of course).

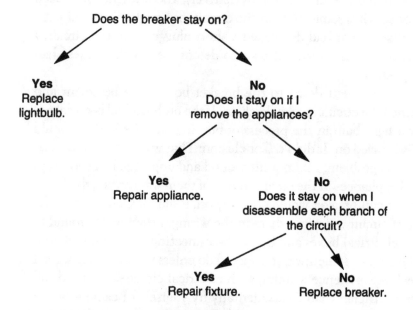

It is obvious that a system of logic like this would be very easy to reduce to a set of fixed rules and hence to a set of instructions to a computer program, or algorithm. This is the essence of an expert system. In fact, medicine and automobile repair are just two of the many areas for which expert systems have been developed. They are enormously useful in dealing with routine problems. Many companies, for example, now use them as initial screens in their customer service system. It is much easier to have a machine ask, "Did you plug the computer in?" than to wait to get a human being to ask you the same thing.

Although expert systems were once touted as examples of machine intelligence, it is obvious that they simply represent the accumulation of the knowledge of human beings. And while they are extremely useful in well-defined and bounded areas, expert systems can make mistakes that no ten-year-old human would ever make. For example, in one famous demonstration, a medical diagnosis expert system was fed information about a car that was covered with rust spots. Its final diagnosis was measles!

When I was a postdoctoral student at MIT many years ago, I attended a party whose host was involved in developing one of the first expert systems—a program designed to conduct a psychiatric interview. The idea was that you would type your problems into the machine, which was programmed to give appropriate answers and to carry out an appropriate conversation with you. For example, what follows is an oft-quoted interview between a "patient" and a modern version of this program, which was called ELIZA:

PATIENT: I have been rather depressed lately.

COMPUTER: Are you depressed often?

PATIENT: Yes, almost all the time.

COMPUTER: Why do you say yes, almost all the time?

PATIENT: Right now I feel I could cry.

COMPUTER: Is it because right now you feel like you could cry that you came to me?

At the party, I watched this program function for a while and realized that it was keying on particular words in its input—in this example, you can see it keying on words like *depressed* as well as repeating whole phrases. Being young and something of a wise guy, when my turn came I typed in

MOTHER FATHER LOVE HATE KILL

and hit return. There was a long pause—you could almost hear the gears grinding in the machine (if it had had gears, that is)—and then the answer came back

"Why do you say that just now?"

In the end, expert systems may have a use in doing midlevel analysis in fields like medicine, but they don't really represent the kind of independent analysis that we normally associate with human intelligence.

Artificial Life and Evolutionary Computing

There are programs other than the neural networks discussed in the previous chapter that are now beginning to command the attention of computer scientists. One group, which has started to go under the name of "artificial life,"* involves what are, in effect, sophisticated versions of computer games to study how systems evolve over time. A typical game might start with a large checkerboard of squares on the screen and two different icons (one a triangle and one a circle, for example). The game also includes a set of rules—for example, you may decide that a circle surrounded by triangles during one move "dies" and disappears in the next. The rules might say that if a triangle is on a square with no circles around, the triangle "captures" the surrounding squares on the next move, and so on. The rules are fed into the machine and the progress of the icons is observed. Under one set of rules, one icon—the triangles, for example—may grow and fill the screen. Under another set of rules, or with a different initial configuration, the

---

*Remember my warning about the sloppy use of language in this field.

triangles may disappear entirely, or they may come into some kind of equilibrium with the circles.

If this sort of exercise were described for what it is—an interesting game that can give you some insight into how complex sets of rules govern the evolution of simple systems—I would have no problem with it. The claims actually made, however, tend to be much more grandiose—you often see artificial life described as a process that mimics the evolution of "living things," with the icons taking the role of successive generations of "living things" and the squares taking the role of the "environment." People have even claimed that you can develop things such as "symbiosis" and "diseases" in this kind of computer game.

There may be a few points of similarity between the results of a game like the one I've described and the results of real-life evolution, but the electrons flowing through transistors in the computers don't begin to capture the complexity of living systems. I have no doubt that artificial life programs will find commercial applications soon (if they haven't done so already), but I don't think they have much to teach us about evolution.

Another new departure goes under the general title of evolutionary computing. The strategy of this type of program is quite interesting, because it borrows from modern genetic theory. The idea is this: A problem is posed, and a set of programs (each of which consists of a set of instructions to a computer) is turned loose on it. For example, the problem may be to take some input numbers, crunch them, and produce the highest possible output number. Some of the original programs may contain instructions to add the numbers together, others to multiply them, others to perform more esoteric operations. After the programs have been run, it will be seen that some are more successful than others at producing high numbers, and it's at this point that evolutionary computing starts to get interesting.

Every program that started the competition consisted of lines of code (i.e., instructions to the computer). A successful program, for example, may say, "Take all the input numbers and multiply them together." Lines of code from each "winning" program in the first competition are then swapped around among all the other winning programs. In effect, the lines of code are scrambled and a new set of programs are constructed from the lines of code that won the first round of competition. The resulting "offspring" programs are then allowed to run for a while, winners are chosen, the lines of code are scrambled again, and so on.

The idea behind this procedure is a rather self-conscious analogy to the natural selection that guides evolution. The lines of code are analogous to genes, and the process of swapping lines of code is analogous to the process by which successful living things mate (and mix genes) with other successful living things. In fact, the original name for these programs—"genetic algorithms"—carries with it the mark of its intellectual ancestry in evolutionary theory.

In the end, this process produces a program that is much better at doing the assigned task than any of the original programs were. Evolutionary computing is particularly suited to solving complex problems in which there are many variables and you're not quite sure how to go about getting an optimal solution by varying them all at once. One computer scientist, for example, likened these problems to adjusting knobs on a shower that had eighty-seven faucets instead of the ordinary two.

The Turing Test

Perhaps the most widely publicized proposal ever made about gauging the intelligence of a machine came from Alan Turing and now goes by the name of the "Turing Test." The basic idea of the Turing Test is very simple. Suppose that you were sitting at a computer console, and suppose that you could communicate with something in another room. This communication might take place through a keyboard and display screen, for example, or it might take place by voice. Suppose you were able to talk for as long as you wanted and ask as many questions on as many different subjects as you wanted. Suppose that at the end of this conversation you were asked to decide whether you had been talking to a human being or a computer. If you could not tell, or if you said that you were talking to a human being and you were actually talking to a computer, then the computer in the other room would be said to have passed the Turing Test.*

There is some dispute over whether Turing actually thought that machines would ever get to the point where it might be feasible to perform such tests. My reading of an article he wrote in 1950 is that he did. But whatever he believed back then, the growth of computing power has been so fast that there are now serious contests to see whether machines can pass something like a Turing Test.

Part of the motivation for the contest is the $100,000 associated with the Loebner Prize, which will be given to the first machine to pass a full-blown Turing Test. We are a long way from that at the moment, so a smaller $1,500 award has been established to mark steps along the way. The general format for these tests is that a group of human judges communicate with machines or other humans through keyboards.

If you read through transcripts of these tests, it's hard not to experience a sense of letdown. Typically, the subject matter of the test has to be very limited—for example, in one recent test the judges were allowed to talk only about wines. Judges are also ordered not to use "unnatural trickery or guile" in their questioning, a restriction that seems to me to make the whole contest pointless. Even so, anyone who mistook a computer in these tests for a human being would have to be very gullible indeed.

But the fact that computers can't pass even a limited Turing

---

*As a matter of historical fact, Turing's first proposal involved a somewhat more complicated three-way interaction between two people and a computer, but the basic idea of seeing whether a human judge can tell the difference between a person and a machine is the same.

Test really isn't the point. It's always dangerous to base arguments on what machines can't do now—it puts you at the mercy of smart technicians and engineers. No matter where computers are at present, it is at least possible to conceive of a computer that could pass the Turing Test. What then? If a machine passes the test, does that mean that we have to assign intelligence or even consciousness to it?

The Chinese Room

This is a problem that was tackled by University of California philosopher John Searle. His argument, which has achieved the level of folklore among the consciousness community, is called the "Chinese Room."

Here is how it works: You are sitting in a room and someone passes you a series of questions written in Chinese (or Albanian or Basque or any other language you can't understand). You then have a set of reference books saying, in effect, that if you get a particular set of characters as input, you are to send a specified set of characters out. Searle pointed out that if those sets of instructions were written well enough, it is quite possible that you could sit in that room, take in the written question, look up the appropriate response in your references, and send appropriate answers back out *even if you didn't understand a word of the question or the answer.* Searle's conclusion (a valid one, I think) is that having a machine capable of passing the Turing Test in no way implies that the machine is either intelligent or conscious. The point of the exercise is that you can put yourself inside the Chinese Room in a way that you can't put yourself inside a complex computer program (or the mind of another human being). You know that the person in the Chinese Room is not aware of what he or she is doing while the test is being passed. Because of this, you realize that passing a Turing Test in no way guarantees that the computer is any more aware of what it's doing than the person in the Chinese Room.

There have been many objections raised to the Chinese Room argument—indeed, there is at least one book that I am aware of devoted to nothing but arguments and counterarguments about this subject. Let me talk about a few of these arguments, just to give you an idea how they go.

The first class of arguments comes from people who really should know better and deals with the question of whether it is actually possible to build a Chinese Room. For example, Frank Tipler, in his book *The Physics of Immortality* (Doubleday, 1994), argues that Searle's example makes no sense because no one could possibly lift books down off a shelf fast enough to make reasonable responses to input questions in real time.

To be honest, I am embarrassed that a fellow theoretical physicist would make an argument like this, for the simple reason that so much of theoretical physics depends on what are called *thought experiments*. These are experiments that cannot be carried out in reality but whose results can lead you to important conclusions. For example, Albert Einstein is supposed to have gotten the idea for the theory of relativity while he was riding in a tram in Bern and looking at a clock tower. He realized that if the tram were to move away from the clock tower at the speed of light, it would appear to him that the clock had stopped. From this, he concluded that it was reasonable to investigate the idea that time may depend on the state of motion of the observer, and from this the entire theory of relativity eventually followed.

Now there are many objections that can be made (and indeed were made) to the theory of relativity. All of these objections were answered in the only way that physicists know how to answer them—by putting them to the experimental test. One objection that was never made (and never *should* have been made) would have been to remark that it is impossible to get a Swiss streetcar to move at the speed of light. That simply is not relevant to the argument. I would suggest that Tipler's counter to the Chinese Room argument falls into that same category.

Daniel Dennett, in his book *Consciousness Explained* (Little, Brown, 1991), gives a somewhat more sophisticated version of this sort of argument. He says, in essence, that you couldn't write down all the possible sentences in Chinese, but would have to have some sort of computer program that did grammar and logical parsing of the incoming words. This program would have to be so complex, Dennett argues, that you couldn't say *a priori* that it wasn't conscious.

Now there is no doubt that if you were actually to set out to build a Chinese Room you would be forced to proceed in this way. Furthermore, it is very likely, as Dennett argues, that the complexity of the grammatical rules and logical sequences that you would be required to include in your machine would make it impossible for you to make the categorical statement that the machine was not intelligent and/or conscious. But the point is that Searle is *not* proposing to build a Chinese Room, any more than Albert Einstein was proposing to put warp drive into a Swiss streetcar. The essence of a thought experiment is that it lays open the logic of a particular problem so that you can see it. It is not necessary that you actually carry out the experiment (although a number of things that were originally considered thought experiments have actually been done). The lesson of the Chinese Room, it seems to me, is that even if a machine passes the Turing Test, it may not have the attributes that we normally expect when we use words like *intelligence* and *consciousness*.

Finally, there is a set of arguments that hold, in effect, that although no single component of the Chinese Room—the person, the books, the input and output devices—is conscious or intelligent, there is a sense in which the entire collection is. It seems to me that this argument might pass muster for a complicated system where it's easy to lose track of the working parts. The virtue of the Chinese Room is that it allows you to get inside the machine, to understand what's going on in a way that's impossible when you confront a computer (or another human being). You *know* that what you do when you carry on a conversation has no relation whatsoever to picking responses from a ready-made list. (If you doubt this statement, go back to the argument about the grandmother cell on page 96 and ask whether there are enough neurons in your brain to hold all possible English sentences.) Because of its simplicity, the Chinese Room allows you to see that something may appear to be making intelligent responses to verbal input when, in fact, it is doing something else entirely.

As was the case with the chess-playing machine, we see that a machine that passed the Turing Test could do so by using a process quite different from those that go on in the brain. Even though we don't understand in detail how the brain works, we can see a pattern developing—a pattern that indicates that even when computers and the brain perform the same function, they do it in different ways. And if this is true, it becomes possible to question one of the leading paradigms of our modern age—the notion that the brain is, at bottom, simply a complicated form of digital computer.

10

Why the Brain Is Not a Computer

The reader must accept as a fact that digital computers can be constructed . . . and that they can in fact mimic the actions of a human computer very closely.

—ALAN TURING

A brain does not look even a little bit like a general-purpose computer.

—FRANCIS CRICK
THE ASTONISHING HYPOTHESIS

It is tempting to look at the human nervous system and see the brain as a kind of digital central processor, with the nerves of the peripheral nervous system acting as input and output channels.

Although many computer scientists have moved beyond this simple view, it remains, I think, the reigning paradigm among popular writers on the subject. It has a lot going for it—it's simple, graphic, and easy to understand. Unfortunately, it is also completely wrong.

In this chapter, I want to explore all the ways in which the conventional wisdom is wrong—all the ways in which the brain is *not* like a computer. In a sense, it's too bad that things turn out this way. It would be pleasant if we could find a simple mechanical analogue to the brain. The result of this chapter reminds me of something I always include in my lecture to freshman students entering the university. There is a tendency in our culture to reduce every issue to simple slogans that can fit onto a bumper sticker. "There is only one bumper sticker I'll allow you to have," I tell the students. "It reads, 'It's not that simple.'" I don't care what the issue is, this bumper sticker will describe it. And, as I hope you'll agree when we get through this chapter, the question of the nature of the brain is no exception.

From any objective point of view, there is no reason at all why anyone should suspect that the brain and the digital computer could be similar in any but the most superficial ways. It is only a slight exaggeration to say, in fact, that the brain is no more like a computer than it is like a bicycle. Nevertheless, the statement that "the brain is just a computer" has been made so often, and has been drummed into the general consciousness so thoroughly, that it is necessary to go into detail to explain why the analogy fails. We can't really make progress in our quest for human uniqueness unless we dispose of this particular canard.

I suppose the real intent of this chapter is to persuade you that if computer people in the 1950s had understood the functioning of the brain, they would not have compared it to a computer in the first place and that we wouldn't have any misconceptions about the connection. Such is the power of an accepted metaphor, though, that it is no longer possible to go back to that pristine state of innocence. Most educated people have been told that because the brain can do computation, it must be a computer, so the burden of proof, rightly or wrongly, is on those who want to argue the opposite.

Before getting into this subject, I want to clarify one point. As we saw in chapter 1, there is a school of thought (I called it mysterian) holding that there is something about human mental activity that must remain forever outside the realm of science. One group within this school argues that the brain cannot be understood by the ordinary laws of physics and chemistry. Arguing that the brain is not a computer, as I do in these next two chapters, in no way implies that the brain is not a material system governed by ordinary natural laws. A bicycle, after all, is not a computer, but it is most certainly governed by those laws. These chapters are simply devoted to advancing the argument that the brain is not a particular type of machine.

The question of the brain-computer analogy naturally breaks down into two parts: (1) Is the brain structurally like a computer, and (2) Can a computer function like the brain? Let me give you an analogy to make this clear.

Suppose someone saw an oxcart going down the road and a plane taxiing down a runway and argued, "They both move on wheels, so they must be the same thing." How would you counter the argument? One way would be to point out all the structural differences between the two—the plane has wings, the cart doesn't; the plane has engines, the cart has oxen; and so on. This would be an argument from structure. Another strategy would be to wait until the plane takes off, then point out that there is something (flying) that the plane can do but the oxcart can't. This would be an argument from function. As far as the brain-computer problem is concerned, I'll be talking about the argument from structure in this chapter and the argument from function in the next.

In principle, the argument from function does not hinge on the argument from structure. Think about the ancient problem of human flight. There have been two historical approaches to this problem. One was to look at the way that things fly in nature and try to mimic it. The fanciful designs of Leonardo da Vinci (as well as many actual machines built toward the end of the nineteenth century) assumed that in order to fly, humans would have to follow the examples produced by natural selection. Until quite recently, however, when advances in material sciences allowed the construction of machines that could remain airborne while powered only by human muscles, this approach yielded little in the way of success. Instead, other approaches, totally different from those developed by nature, put human beings into the air. Think of a blimp and a 747. Neither flies the way a bird does, yet each indubitably flies. In the same way, it is quite possible to imagine that we could build a machine that could do everything that a brain does, but doesn't resemble the brain in structure.

It's important to realize that the argument from structure can never provide an absolute proof. Take the oxcart-airplane analogy as an example. I might start by saying, "The plane has rubber tires, the oxcart has wooden ones," to which you could reply, "Yes, but I could build an oxcart with rubber tires." I might then go on and say, "But the plane has an electrical system, the oxcart doesn't," to which you might reply, "Well, that's true of oxcarts today, but you *could* build one with an electrical system," and so on. When I've run the arguments in this chapter past my colleagues (particularly computer scientists), the discussion seems to bog down quickly in this sort of dialogue—I call it the "putting the tires on the oxcart" discussion. So at the risk of repeating myself, let me say once more that the purpose of the following argument is to establish the possibility in your mind that the human brain may not be much like a computer after all.

The Brain Does Not Operate on the Same Time Scale as a Computer

The neuron operates on the time scale of milliseconds—that is, it typically takes at least a few milliseconds for a neuron to fire, for a nerve signal to travel along its axon, and for the system to clear so that it can fire again. An ordinary transistor like the one in your personal computer, on the other hand, can turn on and off in a billionth of a second (i.e., a million times faster than neurons), and experimental models can turn on and off a thousand times faster than that.

All this talk of milliseconds and billionths of a second may not have much impact on you, so let me give you a simple example of what it means for one thing to be a million times faster than another. Suppose you had one person who could do a particular job in a day, and someone else who took a million times longer to do it. If the first person had started on the job twenty-four hours ago, he or she would just be finishing now. In order for the slower person to be finishing the job at the same time, he or she would have had to have started the job in around 770 B.C. That's how much faster even an ordinary transistor is than a neuron!

On the other hand, we know that the brain can work very fast on some tasks. Here's a demonstration: Lift up your head and look around, then tilt your head. When you do this the visual picture that you have of the world stays vertical—it does not tilt as your head does.

This simple operation is so effortless it's easy to lose sight of the fact it constitutes an enormous computational challenge—only very recently have state-of-the-art machines been able to carry it out in real time. This is because the traditional way a computer analyzes a picture is quite different from the way a human brain does. In its simplest incarnation, a computer system for producing visual fields will break a picture down into small units called *pixels*, and then analyze them one by one. In your TV set, for example, each side of the picture is split into 525 divisions, so the entire picture consists of $525 \times 525 = 275{,}625$ pixels. To tilt the picture, the computer would have to analyze and change each pixel, a process that would take it a long time to perform.

The fact that the brain can perform operations like this so

rapidly means that it must have some way of compensating for the slowness of the individual neurons. In fact, as we saw in chapter 5, the brain is composed of many sets of highly specialized individual groups of neurons. This means that the brain operates in what computer people call a massively parallel way. That is, there are many different pieces of the picture being put together simultaneously, so that although each operation proceeds relatively slowly, it doesn't matter.

Whenever we encounter a task that the brain can do better than a computer (and there are many of them), you can be sure that you will find some clever mechanism like this. And although you can probably program computers to mimic these clever tricks (by doing parallel processing, for example), this is not their natural mode of operation. A computer is much better at using blinding speed (rather than cleverness) to overpower problems. And this leads us to the second difference.

The Brain and the Computer Are Good at Different Things

It is a general piece of folk wisdom in the psychological and computational communities that the brain is very good at things that computers have difficulty doing, and that computers are good at doing things people can't. For example, a computer has no problem remembering long lists of random numbers or even all the guests staying at a major national hotel chain next Tuesday. No human being could possibly keep that sort of information in his or her memory—it is precisely because of this inability that we invented writing. On the other hand, a three-year-old child can easily understand grammatical speech and the use of idioms that a computer can't.

This difference in ability was not known in the 1950s, when people first began to think seriously about the power of computers. At that time, scientists really believed that things such as analyzing pictures and sentences would be just as easy for computers to do as performing numerical calculations and keeping track of data. There is a story, probably apocryphal, that Marvin Minsky of MIT, one of the gurus of the artificial intelligence business, assigned a student the problem of developing a computer visual recognition program as a summer project. If it's true, this story indicates that back then people thought it would take no more than a few months to solve a problem that continues to bedevil the best machines and the best minds we have.

It seems to me, in fact, that the more computers develop, the more we see them as being complementary to the human brain. The names of some of the small portable computers that are now around—Notepad, Personal Assistant, etc.—emphasize the idea of the brain and the computer constituting a partnership, each supplying what the other can't. I believe that had this result been known earlier, the metaphor of the brain as a computer probably would never have been born.

The Brain Evolved, the Computer Was Designed

Another crucial difference between the brain and the computer can be seen by looking at how they got to be the way they are. In chapter 7, we talked about the process of evolution and discussed how it could have led to the development of something like the human cerebral cortex. One of the key ideas that came out of the discussion was the realization that evolutionary systems do not have much of a resemblance to systems that were designed by engineers. (I'll remind you, for example, that in the human eye, the "hardware" that begins the process of producing a visual image actually sits in front of the retina, blocking that same incoming light.) Systems that evolve only have to be "good enough" to get by—they don't have to be the best possible.

We do not yet know much about the details of the wiring of the brain, so I cannot point specifically to examples of the "good enough" principle in the design of neuronal circuits in the brain. I think it is reasonable to expect, however, that once we get "under the hood" and begin understanding how these circuits work, we are going to find many such examples. The way the brain functions in humans (or any animal for that matter) is the result of a long historical process, a process that was not designed to produce what we call higher level cognition. It would be astonishing, therefore, if we did not find all sorts of functional differences between the design of the brain and the design of a machine that is supposed to do the same tasks as the brain. The brain, in short, is an example of evolutionary logic, the computer of electronic logic (although the development of evolutionary computing may make this distinction a bit more fuzzy in the future).

The Brain Is a Chemical System, the Computer Is an Electrical One

However fancy the design, however complex the mechanism, the operation of a computer always boils down to one thing—the movement of an electrical charge in a semiconductor. It is, in other words, an electrical system. The brain, on the other hand, like any other living entity, operates on the basis of chemical reactions. In fact, there are several different levels at which the chemical nature of the brain is made manifest. For one thing, electrical signals are transferred from one neuron to the next by a specific neurotransmitter and its receptors. This was discussed in some detail in chapter 5.

The initial identification of the brain with the computer probably had to do with the state of knowledge about neurons in the 1950s, when people were first beginning to think seriously about calculating machines. At that time, the way that signals were transmitted from one neuron to the next was not known. There were two different schools of thought, which can be characterized roughly as the "spark" school and the "soup" school. People in the "spark" school believed that the transmission across the synapse was something like the jumping of an electrical spark across an ordinary contact. In other words, they believed that the transmission of neural signals was essentially electrical in nature. Followers of the "soup" school, who eventually turned out to be right, believed that the transmission of nerve signals was chemical rather than electrical.

It is not hard to see that if you believed that signals were transmitted from one neuron to the next by what is essentially an electrical current, then there should be a clear analogy between the computer and the brain. The analogy isn't so clear once you bring neurotransmitters into the picture.

As I pointed out in chapter 5, the neuron goes through a complex and as yet unknown procedure for deciding whether to fire, but once the decision is reached, the signals travel along the axon according to their own rules. In this sense, the neuron can be thought of as a switch rather like a transistor—it is either on or off. However, this analogy does not bear close scrutiny. For one thing, the use of neurotransmitters to bridge the gap between neurons means that the signal received by the postsynaptic neuron depends on the receptivity of specific receptors in the neuron receiving the signals. Indeed, as I pointed out, a given neurotransmitter can be either excitatory or inhibitory, depending on the type of receptor to which it attaches. There is no analogue for this process in the computer.

A more important aspect of the chemical nature of the brain is that it is connected to the body's second great mode of communication—the endocrine system. The brain, in fact, exists in an ever-changing bath of chemicals, both those created within itself and those manufactured elsewhere in the body.

Furthermore, this chemical bath appears to play a major role in determining whether or not a neuron fires. A set of inputs that would cause a neuron to fire when the chemical bath has one composition might well not do so if the bath had another. Think of the chemical effects as setting a thermostat in the neuron that determines the threshold for firing. Neuropeptides (one form of neurotransmitter) can diffuse from the neuron from which they were emitted and have an effect on others in the immediate vicinity. Glial cells (which, you will recall, are actually the most common cells in the brain, although they are not neurons) also appear to affect the firing of neurons.

When we leave the central nervous system itself, we find that chemical communication is even more important. As we saw in chapter 6, the hypothalamus has a direct link to the pituitary gland, which in turn controls the levels of hormones in the body. These hormones travel through the bloodstream and are known to affect the workings of the brain.

To take one simple example of the way the body can affect the brain, think about what happens when you haven't eaten for a while. The level of sugar in your bloodstream drops and neurons in the hypothalamus detect the change. Signals are sent upward into the brain, and all sorts of complex behavior is organized, the result being that you eat. Shortly thereafter, your blood sugar levels go up, a rise that is again detected by the hypothalamus, and signals go up into the brain indicating that hunger is no longer a problem.

There are other examples of the connection between mind and body. Just think, for example, about the last time you were startled or emotionally upset and try to imagine yourself solving a calculus problem in that state of mind. (I suspect that much of the testing anxiety that is the bane of teachers from kindergarten to graduate school arises because of this sort of connection between the body's hormonal system and the functioning of the cerebral cortex.)

The connection works the other way, too. Mental states can have a profound effect on the body. Anyone who has a phobia knows this. If I close my eyes and imagine being in an unprotected space high above the ground, for example, it's not long before my palms start to sweat. And anyone who attends operas has seen people sitting still, listening to the music, with tears running down their cheeks. In both cases, purely physical reactions are triggered by the firing of the neurons in the brain, without any external stimuli that could cause them.

The chemical nature of the brain's functioning is summed up by neurophysiologist Antonio Damasio, in his book *Descartes' Error: Emotion, Reason, and the Human Brain*, like this:

> Neural signals give rise to chemical signals, which give rise to other chemical signals, which can alter the function of many cells and tissues (including those in the brain), and alter the regulatory circuits that initiated the cycle itself.

This simple biochemical truth about the human body, incidentally, puts to rest once and for all the notion that there is a mind that sits in our skull and does its thing independent of the body. The brain affects the body, and the body affects the brain—you can't really separate the two. Indeed, some authors like Damasio are beginning to use the term "mind-body" to emphasize this fundamental connectedness.

A Summing Up

Given these fundamental differences between the digital computer and the human brain, it is little short of astonishing that the brain-computer analogy could ever have been made in the first place. As I said at the beginning of this chapter, my purpose here is not to produce a logical proof about the impossibility of a brain-computer analogy, but simply to point out reasons why such an analogy might fail.

But if the analogy doesn't work, what are we to make of all of the developments in computer science, such as neural networks, that seem to be founded on the notion that computers can be built to mimic the workings of the brain?

Here's an example that may help you deal with this question.

Suppose an extraterrestrial came to Earth and observed a large city. Suppose further that this extraterrestrial, for some reason, was absolutely fixated on the notions of traffic and transportation. It would observe that there was every kind of transportation in the city—people driving around in cars, trains and buses running at scheduled times, trucks carrying goods, and so on. The visitor could easily conclude that a city is a transportation system.

Suppose now that the extraterrestrial decides to build an artificial city. It gets a bunch of little robots to drive cars, buses, and trains, and turns them loose. At first, of course, the results don't look anything like traffic patterns in a real city. Then it gets a brilliant idea—"Why don't I look at the way real traffic patterns operate and fix my robots so that they'll mimic it?" So the robots are equipped with something like neural nets, and eventually, to great academic acclaim, the extraterrestrial visitor announces a major breakthrough—his city now has traffic jams at rush hour.

Suppose that, after decades of development, the robots have developed traffic patterns that are, in every way, indistinguishable from those in a real city. Would that model then constitute "artificial urbanization"?

I think most of us can recognize that it wouldn't. Why not? Because, although a city is clearly a transportation system, it isn't *only* a transportation system. In a real city, all sorts of other activities are going on—people elect governments, fall in and out of love, raise families, and so on. All of these activities affect the transportation system, but aren't part of it. In the end, we'd say that it doesn't matter how well you model the transportation system, there's more to the city than that.

In the same way, I would argue that by concentrating on the aspects of the brain that resemble a digital computer, we are missing important aspects of the system—perhaps the most important aspects. The brain can compute, so I suppose it can be called a computer. That doesn't mean, however, that it has to be *just* a computer. It certainly doesn't mean, as I'll argue in the next chapter, that it has to be a standard computer representable by a Turing machine.

There need be no mysticism in this statement. It is possible for the brain to be a physical system, completely describable by physical laws, and still not be a digital computer. After all, as I pointed out earlier, a bicycle is a physical system, completely describable by natural laws, too.

11

Can the Brain Do Something a Computer Can't?

Gödel and Penrose

An engineer, a physicist, and a mathematician were walking down the street when they came to a building on fire. The blaze was getting out of control, and the fire chief ran up to them and asked for help.

The engineer asked to see the blueprints of the building, then gave the chief specific instructions— so many gallons per minute in this window, so many on the roof—and the blaze was soon out. The chief thanked him.

A week later the physicist came to the fire station with a short pamphlet titled General Principles of Fighting Fires and suggested that the *chief incorporate it into his training and operations. The chief thanked him.*

Six months later, the mathematician staggered into the station with a foot-high stack of papers. Unshaven and rumpled, he slammed the stack down on the chief's desk and announced triumphantly "I've done it!"

"What have you done?" asked the startled chief.

"I've proved that fires exist!"

—Anonymous Graduate Student

If I had been asked to try to find a task that the human brain could do but a computer could not, I would probably have started to think about things that involved emotions and feelings—things we normally refer to as "creative" or "artistic." The last place I would have looked is in the field known as the Foundations of Mathematics, a field devoted to dealing with some of the most abstract and rigorous problems in the intellectual world. Yet if we accept the claims of Cambridge University physicist Roger Penrose, it is precisely in this unlikely field that we will find evidence that the brain is fundamentally different from a computer. His books *The Emperor's New Mind* and *Shadows of the Mind* (Oxford University Press, 1989 and 1994, respectively) injected a completely new element into the consciousness debate, an element we will try to deal with in this chapter.

Before we get to the nitty-gritty, though, let me make a confession. Although I have spent much of my career buried in the world of the theoretical physicist, I *hate* doing the kind of formal mathematical reasoning you need to master to understand what this chapter is about. Like most scientists I know, I tend to think intuitively, in pictures, and use formal reasoning to verify (or disprove) what my intuition tells me is right. Doing formal mathematics, to me, is like driving through a traffic jam—I can do it if I have to, but I sure don't enjoy it.

In fact, I can tell you exactly when I realized this. Like many people aiming at a career in theoretical physics, as an undergraduate I did a double major in physics and math. When I went to grad school at Stanford, I figured I'd just go on in the same way and enrolled in a graduate math class.

To understand what happened next, you have to understand something about the state of mathematics today. There was a time when mathematicians devoted themselves to developing tools for calculation—algebra, plane geometry, and calculus are all examples of this, as was the development of more esoteric stuff all through the nineteenth century. The mathematicians who did this work played a crucial role in the development of modern science, and their names litter our advanced textbooks. Since the end of the nineteenth century, however, this venerable partnership has all but vanished as mathematicians disappeared into an abstract world of formal logical systems that had almost nothing to do with modern science. And although physicists occasionally borrow things from that world (as in modern unified field theories, for example), the old easy camaraderie between the two disciplines has vanished.

Of course, as a green graduate student I had no idea that this was the case, so I was completely unprepared for what happened in that class. On the first day, the instructor got up and announced that he was going to prove that a particular equation, one that physicists know describes an electric field in the neighborhood of electrically charged objects, had a solution. I was a little surprised by this, since it was an equation that I (and every other physics student) had solved many times. The instructor then went on to joke, in that "just between us guys" mode you used to get before women started going into science in appreciable numbers, that physicists had their own proof of the existence of a solution. "They say that if you put another electrical charge anywhere near those

first charges, it will only move in one direction with one velocity," he said, barely able to control his glee, "and because of that
they say there must be a solution to the [electric field] *equation,*" at
which point he almost collapsed with laughter.*

I didn't get the joke at the time, but as the class went on I saw
what he was driving at. The problem is that what physicists and
mathematicians mean by the word *proof* is rather different. To me,
the idea that an electric charge moves is proof that there is an
electric field, just as the idea that something falls when I drop it
is proof that there is a gravitational field. To a mathematician,
though, proof means starting from some basic axioms and pro-
ceeding, one logical step at a time, to a conclusion—you may
remember proofs like this from your high school geometry. In the
class I was talking about, for example, the instructor spent ten
weeks of lecture time developing his own version of the proof he
had described.

Now don't get me wrong. I'm not saying this kind of work
isn't important. Somebody has to make sure all the *ts* are crossed
and all the *is* are dotted. I'm not even saying that because something doesn't seem to have immediate practical applications it
shouldn't be pursued. (At the time I was taking this course at
Stanford a pal and I were studying Anglo-Saxon so we could read
the Anglo-Saxon Chronicle in the original. It's hard to think of
anything more useless than that!) It's just that formal mathematical reasoning is one of those hard jobs I'd just as soon leave to
someone else.

You may feel the same way. In addition, there are some aspects of what follows that get a little technical, particularly those
dealing with something known as Gödel's Theorem. Because of
this, I will provide a quick way around, so that readers who don't
want to tackle the details can skip the box on page 168 and keep

---

*I learned later in life that this sort of thing is considered the height of humor
in some academic circles.

following the argument. As for the rest of you, fasten your seat belts.

Gödel's Theorem

In 1900, the great Prussian mathematician David Hilbert addressed a major international conference of mathematicians. In keeping with the meeting's millennial nature, he presented a list of twenty-three major unsolved problems in mathematics. Some of these problems were pretty technical—number thirteen, for example, dealt with the impossibility of solving seventh-degree algebraic equations by using certain kinds of functions. Some were rather vaguely posed—number six, for example, dealt with putting physics on a logical axiomatic basis.* Some of the problems Hilbert posed have since been solved, others not.

The second problem in his list happened to be something that turned the world of mathematics on its head. It sounded innocuous enough—Hilbert wanted to know if the axioms of arithmetic were, in his words, "self-consistent." As the century went on, this came to be defined as a search for a proof that it is always possible, in principle, to find a set of steps or procedures (what mathematicians call an algorithm) that could decide whether any statement in a mathematical system is true or false. The search for such a procedure became known as the "Hilbert Program."

Go back to your high school geometry for an example of what this means. You may recall that plane geometry starts from a set of eleven axioms that are assumed to be true. (For example, "Things equal to the same thing are equal to each other.") In this

---

*I say "vague" because it's hard to know what this would mean for an experimental science like physics, where what is true and obvious can change when new measurements are done. Hilbert probably didn't think about this, though— he was fond of stating, "Physics is too hard to be left to physicists."

system, you can make statements—e.g., "The sum of the angles of a triangle is 180 degrees." There is a procedure you can follow to prove this statement—I can still remember Miss Hawke taking us through it years ago. Hilbert's question concerned the possibility of doing this in systems more complex than plane geometry.

I suppose that if you had asked those mathematicians assembled in solemn convocation a century ago what they thought the answer to Hilbert's question would turn out to be, they would have voted overwhelmingly in the affirmative. After all, what could be more obvious than the proposition that every statement can be shown to be either true or false? One of the biggest (and most obscure) surprises of twentieth-century intellectual history is that things didn't turn out that way.

The first glimmerings of problems, at least as far as the general intellectual world was concerned, showed up in 1902 when British philosopher Bertrand Russell published a paradox that has come to bear his name. There are many ways of stating it, but here's an exercise that will give you a sense of how it works. Suppose you go to your personal library and start looking through every book there. Some books will contain a reference to their own title in their text, others will not. Make a list of those books that do not refer to their own title, then bind that list to make another book. You might title the new book something like *List of Books Whose Titles Don't Appear in the Text*. (Hardly something to get on the best-seller list, is it?) Now here's the question: Should you enter *that* title in the text of your new book?

Well, if you enter that title in the new book, you will have a book that refers to its own title in its text. But the whole point of the book is that it lists only books that don't refer to their titles. This obviously doesn't work. But if you don't enter *List of Books Whose Titles Don't Appear in the Text* in the text of the new book, then the book won't refer to its own title and it should then be added to the list contained in the new book. No matter what you do, you can't. This is what is meant by a paradox.

In 1905, the French mathematician Jules Richard published a similar kind of paradox in arithmetic, which is now called Richard's Paradox. Both Richard's and Russell's paradoxes showed that there are problems with the normal rules of logic, and that these problems seem to arise when you have logical statements that refer back to themselves. The Russell and Richard paradoxes were well-known to mathematicians in the early years of this century, but my sense is that most people who thought about these issues preferred to ignore them in the hope that they would be resolved when the entire Hilbert program was finally carried out.

In 1931, a slight, bespectacled young Viennese named Kurt Gödel published a paper titled, "On Formally Undecidable Propositions of Principia Mathematica and Related Systems," which turned the world of mathematical logic on its head.* In fifteen densely written pages in an obscure journal, *Monthly Publications in Mathematics and Physics*, he showed that the Hilbert program was impossible—that every self-consistent mathematical system of sufficient complexity must have at least one statement that can neither be proved nor disproved. This statement is now usually called the Gödel Statement.

A more detailed description about how Gödel actually carried through his proof is given in the box on pages 168–170, but you don't need to understand the proof to understand what the proof says. The "bumper-sticker" conclusion from Gödel's paper is that any mathematical system of sufficient complexity must either be incomplete or contradictory (by "incomplete" we mean that not all statements can be proved or disproved and by "contradictory" we mean that it is possible to prove both a statement and its opposite). The theorem says, in other words, that every mathematical system that doesn't contain contradictions must contain at least one statement whose truth or falsehood cannot

---

Principia Mathematica (Principles of Mathematics) was the title of a book on mathematical logic that Russell wrote with Alfred North Whitehead.

be determined within the system. Furthermore (and this is the key point for my argument), that unprovable statement is true.

WHAT GÖDEL ACTUALLY DID

The first (and most technically difficult) part of Gödel's paper is devoted to proving that it is possible to assign a number to every proposition that can be stated in a given system. At this point, you may be asking yourself why you can't just write down all the statements and start numbering them. If this is really what you're thinking, it simply illustrates why you and I will never be real mathematicians. You have to *prove* that you can write them down in order, without getting into a situation where one statement may get two different numbers.

In any case, this numbering scheme was very important because it turns out that the Richard Paradox results from a subtle but fundamental confusion about what is meant by a number. It depends, in fact, on the confusion between the meaning of "ten" in the statement, "ten plus two equals twelve," and the meaning of "ten" in the statement, "This is proposition number ten." (Look, are you *sure* you don't want to go back to the main text?)

What Gödel then did was to look at the statement, "This statement cannot be proved"—a statement that asserts its own unprovability. For technical reasons, this statement was cast in this form: "The statement labeled with number so-and-so cannot be proved," with the number adjusted to refer to the statement itself. To make things easier in what follows, let me refer to the statement "This statement cannot be proved," as "Statement A."

In the balance of his paper, Gödel proved the following:

• Statement A can only be proved if Statement not-A

can be proved. In this context, Statement not-A is "This statement can be proved"—the direct contradiction of Statement A. In other words, if Statement A can be proved it leads to a logical contradiction, with both A and not-A being true, and this means that the logical system itself must be contradictory.

- If the system is not to be contradictory, then Statement A is true, even if it cannot be proved in the context of the axioms of the system. (To see why this follows, note that if Statement A weren't true, then it would be possible to prove Statement A and, by the above part of the proof, not-A as well, thereby leading to a contradiction.)

- Therefore, the axioms of the system must be incomplete. There must be at least one statement in the system that can't be proved within that system. There may be more than one, but we know at the very least that Statement A can't be proved.

Having said this, let me make a couple of points. Gödel's work is not what is called a "constructive" proof. Aside from what we have called Statement A, he doesn't tell you how to find unprovable statements or even how to recognize one if you see it. This is important because there are a lot of hypotheses and conjectures in mathematics that everyone thinks are true but no one has ever proved. In the backs of their minds mathematicians working on these statements have to be aware of the possibility that they will never be proved.

An example of this type of statement is the so-called Goldbach Conjecture, which states that every even number can be expressed as the sum of two primes. A prime is a number that can be divided evenly only by itself and one—3 and 17 are both prime numbers, for example. An example of the Goldbach Conjecture in action is the statement 3 + 17 = 20. No one has ever found an even number (like 20) that can't be expressed this way, but no one has been able to prove that we will never find one. Is this because the conjecture is a Gödel Statement? Who knows?

The other point is that Gödel's Theorem, properly understood, is a statement about a specific property of particular types of axiomatic systems. It should *not* be interpreted as an invitation to babble about the end of reason or the need for some sort of cosmic consciousness, as some commentators do.

The Lucas-Penrose Argument

Gödel's Theorem plays a central role in an argument first put forward by Oxford philosopher John Lucas in the 1960s, then amplified and brought to public attention by Roger Penrose in the books cited above. We need to understand that Penrose actually makes two arguments, one of which is discussed here, and the other of which is discussed under the heading of the Penrose Conjecture below. Although these two themes appear and are woven together in the same books, they are in fact logically distinct, and either one could be false without affecting the truth of the other.

The basic premise of this argument rests on the fact that it is possible for human beings to look at a statement and see that it is true, even if Gödel's Theorem tells us that the statement cannot be proved. The only way that a computer can prove or disprove a statement is to follow a set of logical steps from some basic axioms—to carry out the steps of an algorithm. The point of Gödel's Theorem, however, is that there must be a proposition for which this is not possible, a statement whose truth or falsity cannot be decided by arguing in logical steps from the axioms. Thus, there has to be a statement whose truth or falsity can be determined by the human brain, but that cannot be determined by a Turing machine operating an algorithm.

If we accept this argument, then it is obvious that the human brain cannot be a computer. This is what we called the "argument from function" in chapter 10. Penrose, in fact, uses the argument primarily as a way of arguing against what is usually called the "strong AI" (for "Strong Artificial Intelligence") point of view. This view holds that the brain is a digital computer representable by a Turing machine and the mind is a program or algorithm running on that computer. The strong AI position obviously cannot be sustained if there is something that the brain can do that a Turing machine can't. The Lucas-Penrose argument, then, strikes at the very core of the mechanistic, computer-driven view of human intelligence and consciousness.

As you might expect, opposition to this argument has not been slow in crystallizing. In *Shadows of the Mind*, in fact, Penrose presents excruciatingly detailed rebuttals to no fewer than twenty objections to his original thesis, and rebuttals to these rebuttals are surely in the works.

Many of the objections center on the question of how a human can know something that can't be proved. On the purely formal level, for example, you can argue that when we judge the truth or falsity of a Gödel Statement we are, in fact, stepping outside of a logical system and looking back at it—philosophers call such a procedure *meta-mathematical*. Why, it is asked, can't a computer do the same thing?

It seems to me that this sort of objection begs the question. It assumes, in essence, that the process by which the brain decides the truth or falsity of a Gödel Statement is an algorithm embedded in a larger frame of logic than that used by the computer. The point of the Lucas-Penrose argument, however, is that you can't know that. In any case, you can't prove that the brain operates on algorithms by assuming that it does.

Another class of objections has to do with the notion that the brain doesn't know that the Gödel Statement is true or false, but just guesses that it is. You can program a computer to guess, too, the argument goes, so the distinction between the two vanishes.

This objection is more subtle, because it gets to the core question of what it means for a human being to know something—a question that, I'm sure you will appreciate, has a long and honorable philosophical pedigree. Penrose points out that in this context, however, although the computer may indeed be able to guess at the truth or falsity of a statement, it can't know if the guess is correct until a human tells it. But then, of course, you're back to the question of how the human knows. And so it goes.

I'm not sure that scholars will come to agreement on this issue in the foreseeable future, because to resolve it would require an understanding of the brain functions associated with the act of knowing. Nevertheless, at the end of the day, it seems to me that a case can be made that the Lucas-Penrose argument does exactly what it sets out to do. It shows that there is one operation (in this case discerning the truth or falsity of a Gödel Statement) that the human brain can carry out and a digital computer cannot. From this, it follows that the brain cannot be a digital computer.

You should note that it's not necessary to show that *every* Gödel Statement can be seen to be true by human beings. The logic of the situation is such that if we can find even one example of such a statement in any logical system whatsoever, it suffices to prove that the brain can do something that a computer can't, so the two must be different.

Having said this, I must make one more subtle point. In this discussion, I've been using the words *computer* and *representable by a Turing machine* more or less interchangeably. (A Turing machine, you will recall, was described on page 120 as a hypothetical device that changes one bit of information at a time on a tape according to a fixed set of instructions, or program.) This sort of device would prove statements by following a logical sequence, or algorithm, and therefore would clearly have the same limitations as any logical system. The point of the Lucas-Penrose argument is that a Turing machine could not determine the truth or falsity of a Gödel Statement because the only tools at its disposal are those of logic.

It is possible, however, to imagine a non-Turing computer. For example, you might have a machine in which random noise or cosmic rays or some other unpredictable event was allowed to jostle the apparatus and change the instructions from time to time. The operation of such a machine could not be predicted, of course, but the Lucas-Penrose argument wouldn't necessarily apply to it. Given the fact that the brain is a chemical system that exists in a sea of molecules flooding in from other parts of the body, and given the fact that these molecules can and do alter the brain's operation, then the notion of the brain as a non-Turing computer makes some sense. Such a machine would not, of course, be subject to the Lucas-Penrose argument—a point to which we'll return later.

But in the end, it doesn't seem to me that the Lucas-Penrose argument really gets to the heart of the difference between the brain and a standard computer. While it has the advantage of logical precision, it seems to me to miss the central things we normally think of as uniquely human. Let me tell you about an experience I had once that goes a little further in that direction. This was an incident that occurred when I was courting my wife, many years ago. We were in a restaurant in Chicago, and when I looked at her across the table I knew, with more certainty than I've ever known anything in physics or mathematics, that I was in love with this woman. (The exact thought that ran through my mind, as I recall, was "Oh no, not again!") You'll forgive me if I say that the AI guys are going to have to work very hard to convince me that an algorithm running on a Turing machine is ever going to know anything in quite that way.

The Penrose Conjecture

Having established (to some people's satisfaction, at least) that the brain is not a computer, Penrose goes on to make a suggestion about why the difference exists. His basic thesis is that we can't understand the brain using the science at our disposal now but have to develop a branch of science related to the fundamental nature of quantum mechanics. Let me call this the Penrose Conjecture.

Before going into the details of the Conjecture, let me make a couple of points. First, despite being woven together in the same books, the Penrose Conjecture and the Lucas-Penrose argument are not connected to each other. The Conjecture could be wrong, in other words, and the brain could still not be a computer. Second, the Penrose Conjecture involves thinking about two of the great unsolved problems in theoretical physics—the connection between quantum mechanics and the large-scale world on one hand, and unified field theories on the other. I'm obviously not going to have space to go into either of these in any detail here, but both subjects are dealt with in any number of other books, including some of my own.*

When a physicist wants to discuss ordinary-size objects, he or she uses what is called classical Newtonian mechanics. If you think of collisions of billiard balls, you have a pretty clear notion of how the Newtonian world is pictured. Things are described in terms of force, mass, and velocity, and precise measurements and predictions of future events are possible. In addition, in the Newtonian world it is possible to measure something about an object—its position, for example—without changing the object being measured. You can use Newtonian mechanics to describe anything from a galaxy to an invisible particle of smoke in a hazy room.

---

*They can both be found, for example, in the second edition of my book *From Atoms to Quarks* (New York: Doubleday, 1994).

When a physicist wants to talk about an atom, however, he or she uses a completely different branch of science, quantum mechanics. The key difference in this world is that the act of measurement changes the object being measured. Measuring something like the position of an object in the quantum world is like determining the position of a car in a tunnel by sending another car into the tunnel and listening for the crash. You could certainly carry out this measurement, but in the end you could not assume that the car in the tunnel was the same after the measurement as it was before. Because of this fundamental difference between the world of the atom and our everyday world, in quantum mechanics things like electrons are described in terms of quantities called *wave functions*, and the language used is associated with probability rather than certainty. You would use quantum mechanics to describe anything from an electron to a large molecule.

Penrose's main point is that the operation of the brain depends on the kind of science that describes the intermediate world between the purely Newtonian and the purely quantum mechanical. The Penrose Conjecture can actually be thought of as coming in three parts. The first part is that the true explanation of the functioning of the brain is somehow tied up with the (as yet unknown) physics of this intermediate region. The second part involves a guess about how this link will be made. He argues that a fully unified field theory—what physicists call a TOE (Theory of Everything)—will allow us to move smoothly and naturally from the Newtonian to the quantum world. In particular, he suggests that when physicists finally succeed in understanding the force of gravity in the same way that they understand other forces in nature, the resulting theory will naturally bridge the gap. Finally, in the third part of the Conjecture, he argues that a particular structure in cells, called a *microtubule*, is the place where the effects of this new science will play themselves out.

This is a breathtaking set of suggestions, spanning everything from unified field theories to cell biology. I have to confess to a degree of skepticism about this program, if only because I have too much a sense of nature's cussedness to believe that the solution to one mystery (quantum mechanics) will also solve another (consciousness). But the Penrose Conjecture is clearly stated and can be tested. The theoretical neck is firmly on the experimental chopping block, and we'll just have to wait and see what happens.

Why the Penrose Conjecture Doesn't Solve Our Problem

But let's suppose for the moment that the Penrose Conjecture turns out to be completely valid. Let's suppose that (1) the brain is indeed not a digital computer, and (2) the reason is that the brain works according to the rules of a new kind of science located at the nexus of classical physics, quantum mechanics, and unified field theories. We still won't have solved the problem of human uniqueness!

To see this, think for a moment about what would happen once the unified field theories were written down and we could proceed with confidence into the gap between quantum and classical physics. We would then, if Penrose is correct, be able to understand the working of the brain at the molecular and cellular levels.

And then what? Most likely, we would still be able to see the brain as a machine, operating according to known laws of nature. It's just that the machine wouldn't be a digital computer. It would be something else, something as yet unimagined, operating according to laws of nature we have not yet learned.

And then what? If I know anything about human beings, it's this: Once we figure out how something works, some smart engineer is going to come along and figure out how to build something that makes money using that knowledge. Once we understand the brain in terms of Penrose's new science, it seems very likely to me that someone will figure out how to make a new kind of machine—a meta-computer, if you will—that operates according to the rules of that new science. Just as a digital computer operates according to the laws of quantum mechanics, the meta-computer will operate according to the laws of the meta-science.

So in the end, we'll be right back where we are now. We will still have our boundary between humans and animals, but instead of worrying about the boundary on the other side being set by a Turing machine, we'll worry about it being set by a meta-computer. All we'll have done, in fact, is bought ourselves a few decades' respite while the details of the new challenge are worked out.

12

The Problem of
Consciousness

So now we come to a central issue. If the brain is really a physical system, can we ever make a machine that can duplicate or surpass its functions? Can we, in other words, build a machine that is intelligent or conscious or self-aware, as we are?

Before we discuss this question, let me make a comment about the words being used. When we discussed animal intelligence in chapter 3, we agreed to use the term "intelligence" in a loose, colloquial way and concentrate on how animals actually behave. In effect, we said, "Here's what animal X can do—you decide if that makes animal X intelligent or not." I propose to use the same approach for the discussion of "consciousness." I will try to stick to descriptions of capabilities and leave the labeling to you. It's the only way I've found to keep things from getting bogged down in semantics.

Let me start our discussion of consciousness by reminding you of the notion of the "neurological program" that I introduced in chapter 6. This was a hypothetical program in which each mental experience, from seeing grandma on her Harley-Davidson to working a calculus problem, would be correlated with specific neurons firing in specific patterns in the brain.

Suppose that the neurological program has been completed and that you have a book (or more likely, a computer database) that would say things like, "When you see the color blue in this part of the visual field, neuron number 1,472,999,321 is firing, along with . . ." For the sake of argument, assume that you have a list that gives a similar description for every mental experience, or at least for a very large number of them.

The problem of consciousness can then be stated in a very simple form: What is the connection between the firing of those neurons, my experience of seeing blue (or whatever the experience is), and my consciousness of seeing blue? When I see the color blue, or when I see grandma on her Harley-Davidson, I am not aware of the firing of neurons. The experience of those particular visual images (and any other experience you might want to think about) seems to me to be qualitatively different from the firing of neurons. How can we go from a purely physical-chemical system such as the brain to something nonphysical such as our mental experience? What, in other words, is the connection between the firing of neuron 1,472,999,321 and my *experience* of seeing blue?

In passing, I should note that the way we answer this question will affect the way we approach the problems of both machine and animal consciousness. As we saw in our discussion of the Chinese Room in chapter 10, the fact that a machine acts as if it had consciousness does not guarantee that it does. What would a machine have to do to be labeled "conscious"? For that matter, what would it take for us to call a chimpanzee "conscious"? A lobster? A sea anemone? Until we come to some understanding of this problem as applied to the human brain, we won't be able to deal with it for machines or for other animals either.

I Think, Therefore I Am

Every student of philosophy remembers this famous statement made by René Descartes. You will recall that it is the result of Descartes' quest to find something in the world that could not be doubted. He anchored his philosophical system on the bedrock reality of his own thoughts. For our purposes, the crucial aspect of the Cartesian world view was the notion that there was a clear distinction between the physical body (including the brain) on one hand and the nonphysical mind on the other. This so-called mind-body dualism has played a major role in thinking about mental activity ever since Descartes. Philosophers have, in fact, written long and detailed critiques of the Cartesian approach to the world. Certainly the kind of disjunction between mind and body that is implicit in the Cartesian framework doesn't match very well with what we now know about the brain. Nonetheless, there is a sense in which something like Descartes' procedure remains valid for the question of human consciousness.

No matter how my brain works, no matter how much interplay there is between my brain and my body, one single fact remains. For whatever reason, by whatever process, I am aware of a self that looks out at the world from somewhere inside my skull. I would suggest to you that this is not simply an observation, but the central datum with which every theory of consciousness has to grapple. In the end, the theory has to explain how to go from a collection of firing neurons to this essential perception.

Now I am fully aware that none of us can prove that anyone else has the experience I just described. There is an entire school of philosophy, called *solipsism*, based on the notion that the only thing of which we can be sure is our own experience, and that external objects (not to mention other people) simply don't exist. Nevertheless, I think that it is possible to get beyond this inability to provide a tight logical proof. In my view, people who keep insisting on the inability to know about another person's existence are really playing kind of a game—something that might be appropriate for a sophomore bull session or discussion among professors of English, but which really need not detain us very much in the real world. If you don't believe that there is a "you" who sees the world from a position somewhere inside your skull, you might as well quit reading this book now. Nothing I say from this point on is going to make much sense to you. If, however, you are like most people and are willing to agree that you exist and that other people probably do too, then we can proceed.

In terms of this discussion, the problem of consciousness comes down to asking how a system like the human mind and body could produce the perception of self. How, in other words, can a physical system operating by physical laws—laws that we can in principle know and understand—produce the experience of self-awareness that we all share? It is in the answer to this particular question that we find the greatest difference among people who think about the human mind.

A large number of serious scholars have tackled the problem of human consciousness and have produced an extraordinarily subtle and varied landscape of opinion. It would be presumptuous to try to summarize all this thought in a few pages. Instead, I will outline a few points of view that seem to me to be particularly influential in the modern intellectual landscape.

Deniers

One group of thinkers argues, in essence, that the problem of consciousness either cannot or should not be addressed. In its simplest form, this position holds that there is no problem of consciousness at all—that once you understand what the neurons are doing, there's nothing else to explain. Perhaps the most influential of these is the philosopher Daniel Dennett in *Consciousness Explained* (Little, Brown, 1991). Dennett ascribes a kind of romanticism to those who believe that there is something special about human consciousness—something that goes beyond a knowledge of the actions of the physical brain. He makes a clever analogy:

romantic love : love within marriage
as
consciousness needs to be explained : consciousness doesn't need to be explained

(I have to say that I hope he had better luck getting that past his wife than I did!)

Dennett goes into some detail trying to understand the workings of the human brain from the point of view of psychology, particularly the psychology of perception. He discusses at length, for example, experiments on things like the time it takes people to react to the presence of a colored light and draws conclusions about the workings of the brain from the results. He advances what he calls the "multiple drafts" theory of reality—a theory in which the brain is seen as making progressively more detailed views of the external world as information is processed. The idea is that the brain first does a "quick and dirty" analysis of a visual field, then a series of more complex ones, ending with the final full analysis. Each of the intermediate analyses is what Dennett calls a "draft," a fact that explains the name of the theory.

I have no particular problem with this notion. It may be, in fact, that when the neurological program is completed we will find that it is right. It certainly would fit in with what we know about evolution in general and the evolution of the brain in particular. Even if it's wrong, however, it is a valid scientific theory, which can be tested and either verified or falsified. So far so good.

The problem comes when Dennett approaches the problem

of consciousness. The first time I read his book, I became confused because about halfway through I began to think, "Hey—this guy doesn't think that consciousness exists."

This seemed to me to be such a bizarre view that I actually read the book several times, and when that failed to persuade me otherwise I still worried that I was missing something. I'm sure Dennett would deny that this is a proper interpretation of his work, but other scholars (most notably John Searle in the *New York Review of Books*) seem to have come to the same conclusion.

In any case, it is certainly possible to argue that there is no problem of consciousness—that once we understand the neurons, all else is illusion. Let me call this the "Argument from Denial."

My problem with the position comes to this: When a scientist is confronted with data, there are many things that can be done. You can try to fit the data into your theory. You can hope that the data come from incorrect experiments and will be corrected later. You can ignore the data and hope it will go away. Many famous scientists have adopted each of these procedures. The one thing you can't do, however, is to say that the data doesn't exist.

As outlined above, I believe that the most central fact about my existence is I perceive that there is an "I" that observes the world from someplace inside my head. It makes no difference how many details you tell me about the working of my brain and the firing of my neurons. Until you have explained how I come to that central conclusion about my own existence, you have not solved the problem of consciousness. You certainly won't solve the problem by denying that consciousness exists. For me, reading Dennett's book was a little like reading a detailed discussion on the workings of a transmission, only to be told that there is no such thing as a car.

The place where I have most often encountered the Argument from Denial is in talking to neuroscientists. Immersed as they are in the study of the details of neuronal activity, they are apt to brush off questions about consciousness with a wave of the hand and, "Oh, it's just an illusion," then get back to their work. My sense is that they are so focused on working out exactly what the neurons are doing that they just don't want to think about what problems might come next. The more philosophically minded members of the fraternity, however, will acknowledge that there is a problem to be addressed. That's all I ask.

Mysterians

There are people, dubbed Mysterians, who feel that the problem of consciousness will never be solved. These people differ from Deniers, however, in that they accept the existence of consciousness. They simply argue that, for one reason or another, it will never be explained.

For example, philosopher David Chalmers at the University of California at Santa Cruz argues that the discussion of mind-body dualism has bogged down because people keep trying to explain consciousness in terms of things such as neurons and other physical systems. He would prefer to make consciousness one of the basic (but undefined) properties of the universe, something like electrical charge or mass, which form a part of physical theories but are never themselves defined.

A word of explanation. In any physical theory of the universe, there are always some qualities that are measured but not defined. For example, in the standard Newtonian picture, this class includes quantities such as mass, time, and electrical charge. The way they are measured and compared to one another is defined, but they themselves are not defined except in very general and vague terms. They are accepted as fundamental concepts of nature, and all other properties of the universe are explained in terms of them. Chalmers' idea is that the elemental fact of consciousness should join these particular concepts.

This argument, it seems to me, fails to recognize that as knowledge advances, things that were once undefined and "elemental" become defined in terms of more elemental quantities. For example, in the Theories of Everything we talked about in chapter 12, the masses of the various particles are not taken as elemental, but are calculated in terms of quantities more elemental still. Thus, what is a fundamental property of the universe at one level of explanation often becomes something derived at another. There is no reason to suppose that consciousness is any different, or that it is in any way fundamentally undefinable.

My second objection to this approach is rather more subjective. I just think it is too early in the consciousness game to give up. It seems to me that Chalmers' strategy corresponds to forfeiting a football game after the opening kickoff.

Others have proposed more esoteric arguments about the fundamental unknowability of consciousness. For example, philosopher Colin McGinn of Rutgers University has suggested, on the basis of an argument from evolutionary theory, that the human mind is simply not equipped to deal with this particular problem. His basic argument is that nothing in evolution has ever required the human mind to be able to deal with the operation of the human brain. Consequently, the argument goes, although we may be able to pose the problem of consciousness, our brains have not developed to the point where we can hope to solve it.

The problem is that this argument could have been made in the nineteenth century about quantum mechanics, in the eighteenth century about electromagnetic theory, and at almost any time in history about almost any type of phenomenon. You could, for example, easily have applied it to molecular genetics, yet we are well on our way not only to understanding it, but to using it to better the human condition in myriad fundamental ways. Why should consciousness be any different?

Besides, as we pointed out in chapter 7, the brain evolved to its present state by a series of steps (I called them "evolutionary switcheroos") in which systems developed to carry out one task turned out to be suited for another. The development of the ability to perform higher mental functions has often been independent of the need to perform them. There never has been a time in human history, for example, when our survival depended on our ability to write music or to dance, yet we seem to be able to handle both of those with some ease.

Finally, there is a more mystical group that argues that in dealing with the human mind, science has simply reached its limit. They see something like a big Stop sign in the universe—a sign that says "This far and no further." When I read these sorts of critiques of the scientific investigation of consciousness, I get the sneaking feeling that these people aren't so much looking for limitations of the scientific method as living in fear that scientists will actually solve the problem of consciousness. It is almost as if they would rather not know the answers than deal with the consequence of those answers being something they find repugnant. I can sympathize with this point of view, but closing your eyes to a problem never solves it.

As I pointed out in chapter 1, my major objection to this school is that *as a scientist* I simply cannot accept that there is any part of the physical universe that cannot be understood and explained by the methods of science. In the end, I may be wrong about this. Nevertheless, if I look at history, I find a kind of steady intellectual progress. I see more and more things that used to be mysteries brought under the domain of rational scientific thought. If I am asked, then, to make a guess about what will happen in the frontier of consciousness, I find myself in the position of someone watching a horse race and being asked whether or not the horse that has won every other race it has ever been in is the one to back. You may not be able to prove he'll win the next race, but you would be a damn fool not to bet on him.

Materialists

For our purposes, let me define materialism as the belief that the brain is a physical system governed by knowable laws of nature, and that every phenomenon (including mental phenomena) can, ultimately, be explained in this way. I suspect that most scientists today would consider themselves to be materialists. Despite what you might think based on remarks I made earlier in the book, I would put myself in this category as well.

Francis Crick, in his book *The Astonishing Hypothesis* (Simon and Schuster, 1994), gives what is perhaps the most complete and well-thought-out statement of the modern scientific materialists' view of the human brain. The "Astonishing Hypothesis" is this:

> You, your joys and your sorrows, your memories and your ambitions, your sense of personal identity and free will, are in fact no more than the behavior of a vast assembly of nerve cells and their associated molecules. As Lewis Carroll's Alice might have phrased it: "You are nothing but a pack of neurons."

Given this kind of introduction, you would be justified in believing that Crick is an uncompromising materialist of the "The brain is a computer and you are just a machine" school. In fact, to paraphrase physicist Steven Weinberg, Crick is not an uncompromising materialist—he's a comprising materialist. In fact, Crick is located foursquare in an ancient and honorable British tradition—the anticlerical intellectual curmudgeon. He is obviously worried that if people do not accept the Astonishing Hypothesis, they will be driven to accept religion and the existence of the soul.

I'm not sure this is true. I know many people who would balk at the notion of human beings as supermachines but who nonetheless are not practicing members of any religion and who probably don't believe in souls either. Furthermore, as I shall argue later, there are many subtly different interpretations that can be given to the statement "the brain is a physical system." It can easily accommodate the notion that no machine will ever be built that will duplicate the function of the brain. Consequently, when people say they are materialists, we need to find out what kind of materialist they are. Are they the kind who believes that the brain is a machine and our consciousness is simply an illusion? That no noncorporeal entity like the soul exists? That the brain is a computer and the mind an algorithm? All of these positions (and many others) can legitimately be fit under the materialist label.

Does Accepting Materialism Mean We Have to Give Up Human Uniqueness?

I began this book by asking whether there is anything left that is distinctly and uniquely human—whether what we think of as humanity disappears between our new understanding of the ability of animals and our new ability to build computing machines.

We have now seen that it is possible to make a rather clear distinction between the mental abilities of animals and the mental abilities of people. We have also seen that it is possible to argue that there are certain mental functions that cannot be carried out on a standard digital computer. However, as I pointed out in chapter 11, this does not mean that such mental functions could not be carried out on some as yet unbuilt machine in the future.

So now we come to the central question of this book. Given that the brain is a physical system, does it necessarily follow that the brain can be duplicated by a machine? Let me call a research program based on such a replication of the brain the "materialist program," in analogy to the "neurological program" that we defined in chapter 6.

Here's one way you could imagine the materialist program working out: Begin by supposing that we could build an artificial neuron. This artificial neuron would operate according to some as yet unknown laws of chemistry and physics and would include both the electrical and chemical signaling found in the brain. Suppose that this hypothetical artificial neuron could be made to perform all the functions that a real neuron does.

If we can make one artificial neuron, the argument would go, you could then make as many as you liked—even a hundred billion of them. If you then hook these artificial neurons together in a complex net, you could argue that you would have a machine that was the equivalent of a brain, even if it were made out of silicon or something else. It would then be very easy to extend this argument to a machine that had a trillion or a quadrillion neurons—a machine, in other words, that would far outperform the brain. If you were confronted with such a machine, it would be hard to argue that it wasn't intelligent, or even conscious, no matter how you defined those terms. This, I think, is the ultimate dream (or nightmare) of the materialist.

So now let me ask a simple question. Is it possible that the brain could be a physical system, but that we would not be able to carry out the materialist program?

All of the arguments I have made, all of the lines of reasoning I have followed, come together in this single question. I will argue that the answer is yes, that it is very possible that the brain could be a physical system and yet the kind of scenario outlined above will turn out to be impossible. To do this, I first have to present what I think will be the ultimate answer to the problem of consciousness. Once we see this answer, I will then try to show that it is possible to be a materialist as far as the brain is concerned and yet to hope that there is something uniquely human that will never be reproduced by machines.

In order to do this, I'm going to have to do two things. First, I am going to have to talk a little bit about a new kind of science—the science of complexity. I'm going to argue that what we call consciousness is in fact an example of a rather common phenomenon in this kind of science, something called an *emergent property*.

After this groundwork has been laid, I'm going to present two types of arguments to support my conclusion that the materialist program might not succeed. One will be to look at some historical examples of arguments that seemed just as tight and inevitable as this one, but which failed. The purpose of these examples is to challenge the notion that something that seems logically inevitable must necessarily be true. Once this groundwork has been laid, I will lay out a possible (and, I hope, scientifically respectable) scenario in which human uniqueness is preserved.

13

Consciousness and Complexity

Little drops of water
Little grains of sand
Make the mighty ocean
And the pleasant land

—JULIA FLETCHER CARNEY, "LITTLE THINGS"

The Idea of Complexity

Consider a single grain of sand. If it's just sitting on a table in front of you, it's pretty boring, particularly if you consider it *in toto* and ignore the dance of the atoms inside it. Put another grain of sand on the table and the situation doesn't get much better. If you keep adding grains of sand, however, things begin to change. By the time you have a small pile of sand, an invisible web of forces has already begun to operate. Each sand grain pushes on its neighbors and is pulled down by gravity. The net effect of this web is to cancel the forces on each grain, so that none of the grains move.

The more sand grains you pile on, the more complex the web of forces becomes. Eventually, you add one more grain of sand and an avalanche flows down the side of the pile. In other words, the avalanche is a behavior that shows up only when the web of forces reaches a certain level of complexity. If you have to have a million grains of sand before you see an avalanche, you don't get one-millionth of an avalanche in a single grain of sand.

The pile of sand is a simple (perhaps even trivial) example of what has come to be called a *complex system*. A complex system is characterized by having many factors or agents, each of which interacts with other agents. In the case of the sandpile, the agents are the sand grains themselves, and in this simple system each sand grain exerts an influence only through the action of contact forces on its nearest neighbors.

Behaviors, such as avalanches, that appear only when a certain level of complexity is reached, are called emergent properties of a complex system. I want to argue that things such as human consciousness, intelligence, and other higher mental functions are emergent properties of a complex system whose "sand grains" are neurons.

Even in a relatively simple system such as a sandpile, the job of keeping track of the forces on each grain of sand is enormously difficult—it's certainly not the sort of thing you'd want to tackle with just pencil and paper. Only the digital computer, with its enormous ability to store and manipulate information, is capable of carrying out this sort of task. Therefore, the study of complex systems is of fairly recent vintage. What a marvelous irony it would be if the understanding of the brain, which is not a computer, were ultimately achieved through calculations done on the very computers it was supposed to resemble!

Some Necessary Terms

Because the science of complexity is so new, there are a lot of terms that get thrown around—particularly in the popular press—that need to be kept straight. Here are a few that you're likely to run across:

Nonlinear

There is a knob on your stereo system that allows you to control the volume. If you turn the knob a certain number of degrees, you will get a certain volume. If you turn the knob twice that many degrees, you get twice the volume. The response of the system (in this case the output volume) is proportional to the change in the input (in this case the position of the dial). This is called a *linear response*, and when your stereo operates this way it is called a *linear system*.

Most of science before the middle of this century was concerned with the study of linear systems. The reason: The equations that describe linear systems (like the amplifier in your stereo system) are relatively easy to solve. Linear systems, in fact, are the simplest systems we can find in nature and are described by the simplest equations. It should come as no surprise, then, that they were also the systems that scientists understood first.

Go back to your stereo. If you keep turning up the volume, you eventually reach a point where the sound that comes out begins to be distorted. At this point, turning up the knob no longer produces a proportionate response, but something rather different. Instead of a smooth increase in volume, you hear all kinds of squawks and distortions. This is a *nonlinear response* to turning up the dial. When the stereo is operating this way, we say that it is a *nonlinear system*.

There are a lot of systems like this in nature. Think of a rubber band. If you pull on a rubber band with a certain force, it will stretch a certain distance. Double the force and you double the distance. In this regime, the rubber band is a linear system. If, however, you stretch the rubber band too far, it won't come back. It has lost its elasticity, and at this point there is a very different relationship between the amount of force you exert and the amount of stretch that results. The rubber band, then, is another example of a simple nonlinear system.

I mention the rubber band and the stereo as examples of nonlinear systems because there is a common misconception to the effect that physicists did not know about these kinds of systems before the twentieth century. In fact, the theory that describes a rubber band—the so-called theory of elasticity—was begun in the seventeenth century, during the lifetime of Isaac Newton. So although the study of nonlinearity has grown dramatically in recent years, it has an ancient pedigree.

With a few exceptions, the general rule is that exact solutions to nonlinear equations cannot be done by paper and pencil methods but must be obtained using the kind of computational power only available with machines. In the 1950s and 1960s, there used to be roomfulls of engineers and technicians using Marchant calculators, which were basically fancy adding machines, to solve the nonlinear equations that arise in problems such as the design of airplane wings. These calculators were big clunky things with handles you had to turn to carry out the operations.* For all their clatter, these machines yielded, after the expenditure of enormous labor and effort, only approximate solutions to some simple nonlinear equations. The widespread availability of number crunching computers in the 1960s allowed those mechanical machines to be retired and a serious study of nonlinear systems initiated. Today, enormously difficult equations—equations that would have

---

* When I worked at a major national laboratory as a student aide one summer, I can remember that it was considered a sign of great status for one of us lowly assistants to have one of these monsters on our desk.

stumped the best mathematical minds forty years ago—can be solved routinely.

The onset of an avalanche on a sandpile, like the stretching of a rubber band, is clearly a nonlinear effect. Both exhibit sudden changes when certain levels are reached—changes all out of proportion to those that might have occurred previously. In fact, all complex systems like the sandpile are nonlinear, although not all nonlinear systems are complex. It shouldn't be too surprising, then, that the serious study of complexity is also of relatively recent vintage. We just didn't have the computational firepower to think much about this subject until the last decade or so.

Chaos

I don't think any recent fundamental discovery in science and mathematics has been overhyped as much as the phenomenon of chaos. Chaotic systems are nonlinear (although most nonlinear systems are not chaotic). They are characterized by the fact that their evolution in time is exquisitely sensitive to changes in initial conditions. For example, two chips of wood dropped into water on the upstream side of rapids will emerge very far apart on the downstream side. The outcome of the system (the downstream separation) depends sensitively on the initial state (the upstream separation). This is the defining characteristic of a chaotic system.

An example of the way chaotic systems work is the familiar "butterfly effect." The idea is that a butterfly flapping its wings in China, causing a tiny perturbation in the atmosphere, can set into motion a chain of events that will eventually cause thunderstorms in Rio de Janeiro. Whether the atmosphere is really a chaotic system in this sense is, I think, open to discussion. There is, however, no doubt that there are some systems in nature that display this sort of sensitivity and are, therefore, rightly termed chaotic.

One thing I should point out about chaotic systems before we go on is that they are *not* unpredictable. In fact, almost all of our knowledge of chaotic systems comes from computer simulations in which the time evolution of a system is calculated from known equations. If you know the initial conditions of a chaotic system with mathematical certainty, and if you have a computer of infinite capacity, you can predict *exactly* where that system will be at any time in the future. In the real world, of course, such precision in measurement and depth of computing power are not available, so such predictions usually can't be made. Chaotic systems are unpredictable in practice, but they are not unpredictable in principle.

The real significance of the discovery of chaos is this: Until the 1980s, there was an unspoken assumption among scientists that if a system could be described by a simple equation, then its development in time could be calculated. There was, in other words, an assumption that systems that were simple had to be completely predictable. What the discovery of chaos has done is to show that things aren't that simple. Remember my bumper sticker on page 148. It may not, in fact, be possible to make a practical prediction of the future of a chaotic system, even if the system is described by a simple equation.

Complex Adaptive Systems

When I said that the pile of sand grains was a trivial example of a complex system, I had several things in mind. One, which I have already mentioned, is the fact that each sand grain has an impact only on the sand grains nearest to it. Another, perhaps more important, distinction is that once the sand grains take up their positions in the pile, they don't change as more sand grains are added. Not every complex system is like this. For example, if I were making a pile of marshmallows instead of a pile of sand grains, then as I added more on top the marshmallows at the bottom would start to change their shape.

Systems in which the individual agents can change as a result of the activities of the other agents are called *complex adaptive systems.* The quintessential example of a complex adaptive system is the classical Adam Smith market economy, in which each person in the market responds to prices set by others. There is a constant shifting around, and each agent is both affected by and has an effect upon every other agent.

Based on what we already know about the functioning of the brain, it should come as no surprise to learn that scientists consider the brain to be a complex adaptive system. Not only is every neuron connected to thousands of its neighbors by synapses, but, as we pointed out in chapter 11, the emission of neuropeptides causes each neuron to influence and be influenced by neurons to which it has no connection. Furthermore, as we saw in chapter 6, the brain changes according to its experience because synapses are strengthened and weakened as learning and memory proceed. No wonder scientists see understanding the brain as the ultimate challenge in the study of complex adaptive systems.

Is There a Real Science of Complexity?

Because the study of the science of complexity is so new, there are many basic questions to which we do not yet know the answers. One of these—to my mind one of the most important—is the question of whether there are general laws that govern all complex systems, or whether each complex system has to be dealt with on its own terms. There is ample historical precedent for both the "yes" and the "no" answer. There are, in other words, many examples in nature of systems that look very different but obey the same laws, and there are many examples of systems that look the same but are governed by completely different laws.

For example, there could be no phenomena that differ more in their surface appearances than a tropical lake, a star, and a cell. Yet scientists who study these phenomena understand that much of their behavior can be understood in terms of the law that governs energy—what we call the *First Law of Thermodynamics.* It makes no difference whether the energy you are talking about has to do with the fusion of hydrogen into helium (as in a star), the absorption of radiation (as in the lake), or the release of stored chemical energy through burning (as in the cell). All of these processes can be understood as being illustrations that energy can shift from one form to another but can never be created or destroyed. Thus, there is an underlying unity in nature that is not apparent on the surface.

But not all systems are like this. If you look at the shape of a galaxy, a satellite photo of a hurricane, and the cream as you stir it into your coffee, you see the same kind of spiral pattern. It is tempting to suppose that phenomena so similar ought to be caused by the same physical mechanism. In fact, they aren't. Totally different mechanisms operate on a galaxy, a hurricane, and cream in your coffee to produce the same final result. In this case, we have a situation in which different laws give rise to similar phenomena.

So where on this scale do complex systems fall? Is there some generalized "First Law of Complexity" that will describe both the human brain and the Adam Smith marketplace? Or are the two simply different phenomena that share the property of complexity the way a galaxy and a hurricane share the property of being shaped like a spiral? For the record, my guess is that the search for general laws that underlie all complex systems will probably not be successful. In other words, I think that the brain and economic systems will turn out to be much more like hurricanes and galaxies than they are like stars and tropical lakes. For a very eloquent and passionate presentation of the opposite point of view, I recommend Stuart Kauffman's book, *At Home in the Universe* (Oxford University Press, 1994).

Consciousness as an Emergent Property

Consider a single neuron. Although infinitely more complex than a grain of sand, a single neuron can do only a limited number of things. It can generate an action potential, of course, but in the absence of other neurons there is nothing to which that potential can be communicated. A single neuron certainly could not perform a high-level function such as recognizing a predator or solving a calculus problem. In this sense, the single neuron is similar to the grain of sand with which we started this chapter.

Now start adding and connecting neurons one by one. These new neurons, obviously, will give the system the capability of performing new functions. There are two possibilities for the way these capabilities could develop. As more and more neurons are added, new properties could develop gradually. Alternatively, as we saw with the grain of sand, new properties could appear suddenly as emergent phenomenon in a complex system.

We cannot, of course, actually perform an experiment like this. However, it seems reasonable to suppose that if something as simple as a pile of sand can show emergent behavior, so should a collection of neurons. I will take as a working hypothesis, then, that as we add neurons to our nascent brain, we will see the same sort of behavior that we see in any other complex system. When we reach a certain level of complexity, new kinds of phenomena will manifest themselves.

Given the level of complexity of a single neuron and the degree of connectedness of the brain, it also seems to be reasonable to suppose that there would be more than one kind of emergent property that characterizes the system, and that these properties will appear at different levels of complexity. The result will be a sort of cascade of emergent properties as more and more neurons are added to the system.

What I am suggesting, then, is that if we build a collection of neurons, adding one neuron at a time, the system will go through a series of discontinuous jumps, each jump corresponding to a new kind of emergent property—a new "avalanche"—characteristic of its new level of complexity. The kinds of phenomena we refer to as consciousness and intelligence would, in this picture, correspond to emergent properties at the higher levels of the cascade.

It also means that whenever we find a big gap between the mental abilities of one species and those of its nearest relatives, we are probably looking at something like an emergent phenomenon.

I should mention that this pattern of successive discontinuous changes is common in natural systems. For example, there are many steps between smooth flow and full turbulence in the flow of water, with each step corresponding to a sudden onset of a more highly organized and complex flow.

Animal Consciousness

Although we cannot perform the experiment of hooking neurons up one by one in the laboratory, there is a sense in which nature has already performed it for us. Think back to the stroll through the phyla we took in chapter 3. We saw that a relatively simple neural system, such as that possessed by the sea anemone, was capable of developing fairly complex behaviors. While a single neuron cannot recognize a predator and or send signals that result in the organism fleeing from it, a few hundred neurons seem to be able to. I would suggest that this is the first kind of emergent property that we would see if we began connecting neurons. As we keep adding neurons, we find new types of behavior that represent new emergent properties of the system of connected neurons. By the time you get to 500 million, activities such as learning, memory, and fairly detailed and comprehensive analysis of visual fields become possible (I remind you that the octopus can do these sorts of things and 500 million neurons is about the size of its brain).

This picture of the development of the brain actually explains many features of the evolutionary history of the human race. In chapter 2, we argued that there was a specific time—about 2 million years ago—when hominids became, in some sense, "human." If the appearance of *Homo erectus* corresponded to the collection of neurons we call the brain reaching the point where new emergent properties became evident, we can understand how such a sudden change could have occurred.

When we talk about the evolution of consciousness, we should expect the notions of discontinuous change corresponding to increased complexity to play an important role. This notion also tells us that it is quite possible for human beings (who have, after all, the largest and most complex cerebral cortex in the animal kingdom) to be qualitatively different from other animals at the level of mental functioning, even though at the chemical level we are almost identical.

This is important, because in the discussion of animal consciousness it often seems to be taken for granted that there has to be a smooth, continuous dependence of consciousness on brain size. Here, for example, is a quote from Carl Sagan and Anne Druyan's book *Shadows of Forgotten Ancestors*:

> If [a spider's] brain is one millionth the mass of ours, shall we deny it one millionth of our feelings and our consciousness?

With our understanding of the properties of complex systems, we can see that there is no particular reason to believe that a spider is one millionth as conscious as a human being, any more than there is any reason to assume that a grain of sand can exhibit one millionth of an avalanche. Human uniqueness among the animals is a very reasonable outcome of the notion that the brain is a complex adaptive system.

Let me suggest a simple way to visualize different ideas about the evolution of the brain. If there are no emergent properties as brains become more complex, then you can visualize the progression from the sea cucumber to *Homo sapiens* as a kind of smooth ramp. This is basically the assumption that underlies the quotation given above. An evolutionary path in which emergent properties played a role, on the other hand, would look more like a staircase, with sudden changes in mental capacity corresponding to each new emergence.* This particular visualization will help us in the next chapter, when we return to the problem posed in chapter 1.

Machine Consciousness

So can a machine like a computer be conscious? Look at the problem this way: If we go through the process of building up a system of transistors, adding them one by one as we did for the neurons, then we could expect to see emergent properties in that system, just as we saw them in the neurons. Our question then centers around whether or not it is possible to build a machine that has exactly the same sets of emergent properties as evolution developed for the human brain. This is a much sharper definition of the problem we posed in chapter 1, where we asked whether, when all is said and done, there is anything left that is uniquely and distinctly human.

It is important to realize that by posing the question in terms of emergent properties, we are avoiding the necessity of having to go outside the realm of science to find an answer. It may turn out that it is possible to build a machine that is conscious in the sense that a human being is conscious. It may be possible to build a machine that has sets of attributes that many people would de-fine as "conscious" but which is not like the human brain—that is conscious, perhaps, in a different way. It may, on the other hand, turn out to be impossible to build a machine that can approxi-mate consciousness and the human brain at all. I simply want to insist on one thing: *This is an open question.*

---

*Don't picture this as a single stairway leading up to human beings but as a twisted, branching set of stairs driven by the operation of natural selection on animals in many different environments.

What Would a Theory of Consciousness Look Like?

It should be obvious to you that we are a long way from being able to enunciate a complete theory of consciousness—we'll have to solve the binding problem and a lot of other puzzles like it before we get to that point. This doesn't matter much for the question of human uniqueness that we're investigating in this book. All we need to understand is that when a theory of consciousness does emerge, it is likely to involve the new science of complexity, particularly the notion of emergent properties. Nevertheless, for the sake of completeness, we could take a look at some preliminary theories of consciousness to see how people are thinking about this problem.

First you have to realize that when you raise this question you discover that our knowledge of the working of the brain is extremely primitive. As we saw in chapter 6, we haven't progressed very far in our understanding of how the brain accomplishes a comparatively simple task like putting together a visual picture of the world, and the production of consciousness is sure to be more complicated than that. Nevertheless, there are a few beginnings of brain-based theories of consciousness, and I summarize them here to give you a sense of what they look like. (Obviously, I'm not going to be able to do justice to any of them in a few short sentences.)

Many of these theories pay a great deal of attention to the constant back-and-forth flow of information between the brain and the body. For neurobiologist Antonio Damasio, for example, consciousness arises from a constantly renewed interaction between the brain's awareness of the state of the body (information that is conveyed both electrically and chemically) with memory and other higher cognitive functions. The central idea here is that the brain is constantly updating its picture of the state of the entire body, and it is this complex process that is involved in producing consciousness.

For Nobel Laureate Gerald Edelman, consciousness is more

of a function of the brain itself. He suggests that it arises from a complex reciprocal flow of information between groups of neurons he calls "maps." Edelman pays a great deal of attention to the development of the brain and the formation of synapses. Using language that should remind you of the evolutionary process itself, he suggests that groups of neurons are selected as the brain matures. Neurons that aren't selected for this function, Edelman argues, die off and disappear, much as neurons that make the wrong connections commit cellular suicide.

Francis Crick and his colleagues place the origin of consciousness in the high-frequency oscillations of signals that occur in the brain (we discussed these oscillations with respect to vision in chapter 6). For them, the origins of consciousness are to be found in complex ongoing interactions among specific neurons—interactions that we can begin to detect in those oscillations.

All of these theories have been thought through to a level of elaboration that can (and does) take a book-length manuscript to explain fully. Any of them could develop into a theory that embodied the notions of emergent properties I've been sketching. I think all the authors would agree, however, that we are still a long way from a complete theory of consciousness based on our understanding of neural function.

A Word About Words

One of the things we noted in our discussion of intelligence in chapter 3 is that people often have a great deal of difficulty in dealing with common words. When we use a word such as *consciousness*, we all think that we know what we mean. The problem is that we all think it means something different. Since everyone feels that he or she "owns" the word, enormous arguments ensue when people feel that their ownership of the word is being threatened by someone else's usage.

Let me give you one example. I first became interested in the problems of consciousness when I was invited to join the Krasnow Institute for Advanced Studies at George Mason University. A multidisciplinary discussion group was formed to talk about the general problems of consciousness and complex adaptive systems. The problem of "ownership" quickly became evident, so I proposed that we devote one of our discussion periods to an attempt to come to an agreement among ourselves about what we meant when we used various words. My motive in doing so was simply to avoid all the semantic arguments we seemed to be getting into. I prepared a list of words, starting with *brain*, running through *intelligence* and *consciousness*, and ending with *self-awareness*, that seemed to generate a lot of argument and put together a list of definitions to serve as a basis for discussion.

I expected we'd run into trouble, but I had no idea we'd have the kind of trouble we did. Would you believe it if I told you that this group of professors and academics spent two hours in heated discussion, and in the end could not agree on a common definition of the word *brain*, never mind *consciousness* or anything else?

This sort of problem arises because of a strange kind of impoverishment in the English language. We have a single word, such as *intelligence*, that is supposed to cover everything from an octopus to a human to a chess-playing computer like Deep Blue. This just isn't going to work, especially when we start to build machines that we want to call "intelligent" but which we know don't work in the same way as the human brain.

I don't think this problem can be solved, but it can be made less disruptive. My approach (as you saw in chapter 3) is to refrain from using words like *consciousness* in any but the broadest, most generic, sense. Instead, I describe particular systems as accurately as I can and let the audience decide whether the word applies to that particular system. This approach allowed us to get through a fairly complex discussion of animal intelligence without ever forcing us to face the problem of whether a particular animal was *intelligent* or not (or, worse, having to face the problem of defining what "intelligence" is in the abstract).

Let me propose that we use the same approach when we talk about consciousness, whether of animals or machines. We should simply state what the animal or machine can do and then leave it to our audience to decide whether they want to apply the concept of intelligence or consciousness or self-awareness to something that possesses that particular set of attributes.

We can learn a lesson about the wise use of words by thinking about a building sitting out in the Arizona desert called Biosphere II. The original builders of this dome were motivated by the desire to build a closed, self-contained ecosystem—their self-conscious goal was to build a prototype for colonies on the Moon and Mars. The idea was that the building would be an analogue to the Earth, an ecosystem the builders called Biosphere I. Instead of having the wastes of the ecosystem in Biosphere II absorbed by the rest of the planet, however, they are processed by machines in the building's basement. Thus, Biosphere II achieves some of the same ends as Biosphere I, but operates in a different way. Two things are certain, though—no one would ever mistake the overgrown greenhouse for the real Earth, and no one ever got upset about the building's name.

It seems to me that we ought to take a leaf from the Biospherians' page when we talk about concepts such as intelligence and consciousness. Instead of tying ourselves in knots trying to decide whether a machine like Deep Blue is or isn't intelligent, why not just say that human beings carry Intelligence I and Deep Blue has Intelligence II? In this way, we can accommodate the obvious differences between a computer and the human brain while accepting that the machine can do some of the same things the brain does. If we use this scheme, then there's no reason to believe that we won't find Intelligence III, IV, V, and so on.

The same kind of terminology could be used for consciousness. There are other kinds of consciousness that need bear no more resemblance to human consciousness (Consciousness I) than Biosphere II does to the real Earth. And who knows—perhaps eventually we'll be comfortable with assigning machines Consciousness II, III, IV and so on as well.

In this language, the central issue is not whether we can build machines that are conscious or intelligent, but whether we can build machines that exhibit Consciousness and Intelligence I. The language forces us, in fact, to focus on the differences among tasks the human brain can perform and those performed by machines. In the end, this is where we should be putting our efforts anyway.

14

Is There Anything Left for Us?

> *What is man, that thou art mindful of him? and the son of man, that thou visitest him?*
>
> *For thou hast made him a little lower than the angels, and hast crowned him with glory and honour.*
>
> PSALMS 8:4–5

Restating the Problem

Let me begin by reaffirming my belief that the brain is nothing more than a physical system. It may be a highly complicated system, one that involves both electrical and chemical methods of communication. It may be interconnected like nothing else in the universe, but at bottom it is still a system made of atoms and molecules. There is no need to postulate the existence of any-

211

thing like a soul or any special kind of mystical experience to understand the brain or consciousness.

Starting from this position, it seems hard to deny the possibility of carrying out the materialist program—of creating an artificial brain and/or consciousness. From that beginning, as we have already pointed out, it is only a short step (in principle, at least) to making machines that will duplicate everything the brain can do, in which case there would be nothing left that is uniquely human. I have to point out that, although the machines that we can build today are nowhere near this level, we have to consider what would happen if they were built someday. In this scenario, *Homo sapiens* will probably wind up being just a transitory biological stage between animals and the new silicon-based intelligence. Indeed, some commentators go so far as to refer to these supermachines as "life forms," and suggest that they will replace us the way mammals replaced dinosaurs 65 million years ago.

Is there any way to avoid this outcome? I don't have a definitive answer to this question—no one does—but on the surface it certainly seems that we are caught in a tight logical trap.

Nevertheless, there are a number of historical examples of arguments that seemed just as logical and whose predicted outcomes seemed just as inevitable as this one, but which nevertheless failed. Perhaps a look at some of those examples will help us to see how dilemmas like this actually play out. There are several ways such "airtight" arguments can fail, and I'll illustrate each with a historical example. The point of these examples is not to prove that the materialist program must fail, of course, but to show that what appears to be an unresolvable dilemma at one stage of knowledge may well turn out to be wrong or irrelevant at a later stage.

What's important in the following discussion is that while we know the brain is a complex adaptive system, we simply don't know what surprises await us as the science of complexity develops. The historical examples demonstrate ways in which human uniqueness can be maintained even in the presence of extremely advanced computing machines.

The Divine Calculator

Isaac Newton bequeathed a universe of surpassing order and regularity to his successors. Because of his work, many people thought of the universe as a kind of clock—God had wound it up at the Creation, and now it was just ticking along on it own. By studying the motions of the clock, we could understand the mechanisms of the universe and what was in the mind of the Creator when it was made. We could also use Newton's Laws of Motion to predict the motion of physical systems. Not only the majestic orbits of the planets, but paths of comets, the ocean tides, and the formation of the solar system seemed to fall into place in this system.

The paradigm problem of Newtonian physics is the motion of billiard balls on a table. In a typical problem—the sort of problem every freshman physics major learns how to solve—you are told the masses, positions, and velocities of the billiard balls at a given time, and then you are asked to use Newton's Laws to find the velocities and positions of the billiard balls at any time in the future. If the problem is simple enough—if there are not too many billiard balls involved—this can usually be done.

Given this background, it is not too surprising that some of Newton's followers came to believe that nothing was beyond the capabilities of their new science. Here, for example, are the words of Pierre-Simon, Marquis de Laplace, one of the greatest of the Newtonians, in an essay in his book *Théorie analytique des probabilités* in 1812:

> [This] investigation is one that deserves the attention of philosophers in showing how in the final analysis there is a regularity underlying the very things that seem to us to pertain entirely to chance, and in unveiling the hidden but constant causes on which that regularity depends.

Since Laplace was one of the greatest Newtonian scientists,

who gave us, among other things, the foundation of our current theory of the tides and the theory that describes the formation of stars and planetary systems, we can be sure that this sort of thinking represented mainstream ideas. The Newtonian world, then, was one in which nothing was unpredictable, in which everything proceeded according to the operation of known laws.

But what happens to human free will in a universe which is, in effect, a gigantic set of gears? A Newtonian might argue like this: Suppose you knew the position and velocity of every particle in the universe at a given moment. Then, using exactly the same techniques we use for billiard balls, we would be able to calculate the exact position and velocity of any particle in the universe at any time in the future.

Of course, this would be a very difficult calculation, and no human being in Laplace's time (or today, for that matter) could ever hope to carry it out. But suppose you invoke a "Divine Calculator," a being with sufficient calculational abilities to carry it through? Newtonian scientists could imagine that such a being could exist, at least in principle, and that created problems for the notion of human free will.

Here's why: If one of those particles whose future you can calculate happens to be in your right thumb and I can tell you where that particle will be fifteen years from now, it is clear that you do not have the choice of being elsewhere. Thus, there seemed to be a fundamental conflict between the notion of a human being who could choose his or her future course of action and the existence of a deterministic set of equations that describe the motion of any particle in the universe.

Actually, I rather enjoy posing the problem of the Divine Calculator to classes made up of students who are not scientists, because on the surface it's quite troubling. It has, in fact, the same sort of intellectual resonances we find in the conflict between the materialist program and human uniqueness. It seems to be telling you that you have to choose between science and rationality (as they are embodied in Newton's Laws of Motion) and an aspect of human existence we care a great deal about (free will).

But the Divine Calculator turned out to present a false dichotomy because the universe envisioned by the Newtonians is not the universe we live in. As it happens, matter is composed of atoms, which are themselves composed of smaller particles such as electrons and protons. The motion of these particles is not described by Newton's Laws, but by quantum mechanics.* It turns out that the laws of quantum mechanics contain something known as the Heisenberg Uncertainty Principle, which says, in effect, that when you get down to the scale of individual atoms, it is impossible to measure both the position and the velocity of a particle at the same time.

What this means is that a century and a half after scientists could speak confidently of eliminating chance from the universe, Heisenberg discovered that the laws of quantum mechanics rendered the whole question moot. It's not that the old, seemingly airtight, argument was incorrect. It is probably true that *if* you could find the exact position and velocity of every particle in the universe at a given time, you could, in principle, calculate the entire future of the universe. The point is that the Heisenberg principle says that you cannot know the position and velocity of even one particle at a given time, much less the position and velocity of every particle in the universe. The development of quantum mechanics didn't disprove the Newtonian argument, nor did it show that the problem associated with the Divine Calculator was a product of faulty logic. It simply made the whole argument about the Divine Calculator pointless.

What would it take to have our conflict between the materialist program and human uniqueness play out the same way? To see how this might work, note that the statement we are thinking about takes this form: *If* we can analyze a complex system like the brain, then we can reproduce it. Suppose that as the science of

---

*There is no particular reason why atoms *should* be described by Newton's Laws, since the experimental basis of these laws deals only with large-scale objects. This subject is discussed more fully in my book *From Atoms to Quarks*, (New York: Doubleday, 1994).

complexity develops we find that the "if" clause can't be fulfilled, even in principle. Suppose, for example, that once we get past a certain level of complexity it is no longer possible to analyze a system—to keep track of how all the parts fit together. If this happened, then the science of complexity would have developed in such a way that our dilemma becomes as moot as the Divine Calculator.

E. O. Lawrence and the Supercyclotron

In 1932 the American physicist Ernest O. Lawrence, working in temporary shacks behind the physics building at the campus of the University of California at Berkeley, built the world's first cyclotron. The cyclotron is a machine that accelerates a proton (one of the particles that makes up the nucleus of the atom) to high energies and allows it to collide with a target. By studying the debris produced in such collisions, scientists hoped to (and eventually did) uncover the basic structure of the nucleus and the particles that exist inside of it.

The structure of a cyclotron is easy to describe. Its major working parts are two large sets of magnets. They are shaped like what you'd get if you took a circular layer cake and separated the upper and lower layers, so that there is a small space between them, then sliced each level of the layer cake in two, so that you have a split circle on top and a split circle underneath. Each magnet was shaped like a capital D, and each was, in fact, referred to as a Dee. There were four of them—two on top and two underneath.

Protons are introduced into this structure at the geometrical center, between the upper and lower magnets. It is a property of charged particles like the proton that when they are in the presence of a magnetic field they tend to move in circles.* In the

---

*It is this phenomenon, for example, that gives us the Northern Lights. In that case, the magnetic field is supplied by the Earth.

cyclotron the protons move in circles, but each time they get to the place where we "sliced the layer cake," the machine is arranged so that the protons are given a little push. Because of this push, when the proton gets to the other side of the gap between the magnets it is moving a little faster than it was when it entered the gap.

Because of this faster motion, the proton travels in a slightly bigger arc, so that by the time it moves through the 180 degrees to the next gap it is a little bit farther from the center than it was at the beginning. At this time the proton is once again accelerated, swings into a larger arc, gets to the other side, is accelerated, and so on. As a result of these successive accelerations, the proton spirals out from the center, moving faster and faster until it reaches the very edge of the magnets. Here it is allowed to move out in a straight line, more or less as a stone moves out from a sling, until it hits the target. The cyclotron was the first machine that was graced with the title "atom smasher," although this is something of a misnomer. Even a fluorescent light can "smash" atoms, in the sense of tearing them apart. It would have been more correct to call the cyclotron the "nucleus smasher." (Okay, I know this is an academic point. Humor me.)

The first cyclotron Lawrence built was a tiny affair—you could have held it in your hand. It produced protons of much too low an energy to do any serious study of the nucleus. As the 1930s progressed, however, Lawrence's team built bigger and bigger cyclotrons. Their primary technique was to make the magnets bigger and then crank up the acceleration. Although the cyclotron was not the first machine to split a nucleus artificially, it was the workhorse machine of the 1930s, when nuclear physics was first being explored. In fact, Lawrence won the Nobel Prize in 1939 for his development of this machine. (He was the first American associated with a state university to get the prize.) During the late 1930s, he dreamed of building something you can think of as a "supercyclotron."

During World War II Lawrence, like most physicists of his

generation, worked on the Manhattan Project. After the war, however, he returned to this machine. Lawrence felt that the proper way to design the supercyclotron was to do what he had been doing for so long—that is, simply build bigger magnets. In fact, the magnets he designed were to have been over 15 feet across and weigh some 4,000 tons. In these magnets, protons were to be accelerated to the unheard-of energy of 100 million volts.

While Lawrence was out talking to industrial and government big shots about money for his machine, however, theorists began to realize that a then-obscure theory—the special theory of relativity—predicted that it would be impossible for Lawrence to build the machine as planned. As you may know, this theory predicts that when the speed of an object approaches the speed of light, the object begins to get heavier. If you throw this fact into the equations that govern the cyclotron, you find that once the protons have made a certain number of cycles around the machine, the extra mass slows them down, and it takes longer to go around the arcs. Without going into the technical details, this development would have the effect of making it impossible (or at least very difficult) for the cyclotron to accelerate particles to higher energies. In the end, Lawrence's supercyclotron was never built.

So here is another way a seemingly airtight argument can fail. In our format, we can paraphrase Lawrence's argument as, "If I can make a bigger magnet, *then* I can make a supercyclotron." The first part of this statement is unexceptional. We have no problem today making magnets much bigger than Lawrence needed. The problem is that the second part of the statement does not follow from the first, for reasons that no one could have foreseen without knowing about the theory of relativity.

This example illustrates another way that the materialist program could fail. It may turn out that as the science of complexity develops there will be laws that say, in effect, that when you get to a certain level you cannot duplicate systems, even if you understand them completely.

In passing, I should note that what relativity showed was that you could not accelerate particles to high energy using the cyclotron—not that particles could not be accelerated at all. In fact, today we can accelerate particles to much higher energies than Lawrence ever dreamed of by using a machine called a synchroton. Most of the large accelerators that you have heard about are probably of this type. Thus, in this historical example, human ingenuity found a way around a fundamental barrier imposed by nature. It is possible that if the theory of complexity turns out to impose the kinds of limits that I am suggesting, it might also happen that those limits can be circumvented by clever engineers.

Is There a Gödel-Type Theorem Waiting for Us in Complex Systems?

On the face of it, there seems to be nothing more straightforward than the notion that any system, no matter how complex, can be analyzed completely and reproduced. This has been a hidden assumption in almost every discussion of complexity I've ever seen. Most writers tacitly assume that the only inhibiting factor in analyzing complex systems is human ingenuity and sometimes the availability of computing power.

For example, in developing the ideas discussed in chapter 12, philosopher David Chalmers introduces the notion of a "synthetic brain" in which neurons are replaced, one by one, by silicon chips. The idea is to show that there is no place at which you could draw the line between the natural and the synthetic system. The assumption implicit in this argument is that there are no fundamental laws of nature lurking in the background that would make this program unattainable.

In fact, compared to the brain, with its hundred billion highly interconnected neurons, the level of complexity on even the most advanced microchip is negligible. This means that when we begin pushing silicon systems to the level of complexity that we find in the brain, we are making a large extrapolation—a big jump—with no guarantee that the jump can actually be made.

Here again, a historical example illustrates what I mean. What could be more obvious than the statement that every proposition in a mathematical system can be proved to be either true or false? This statement certainly seemed obvious when David Hilbert proposed his famous twenty-three problems in 1900. What actually happened, as we discussed in chapter 11, was that Kurt Gödel proved that when you get to a certain level of richness in the logical system, there would always be statements that can neither be proved nor disproved.

Biologist Jack Cohen and mathematician Ian Stewart, in their book *The Collapse of Chaos: Discovering Simplicity in a Complex World* (Penguin Books, 1994), present a credible scenario about how something like the Gödel Theorem could arise in the study of complexity. As we have seen, it often happens that the properties of a given complex system can be seen as the system is simulated on a computer but cannot (at least for the moment) be predicted in advance. Cohen and Stewart suggest that the phenomenon of emergence may be connected to the existence of propositions in the systems that, although they can be proved, involve proofs so long as to be meaningless to human beings. In their words:

> If we wish to use reductionist rules to explain and understand high level structures, then we have to follow [a] chain of deduction. If that chain becomes too long, our brains begin to lose track of the explanation, and it ceases to be one. This is how emergent phenomena arise.

This is an important (though as yet unproved) suggestion, one that may have an important impact on our discussion of human uniqueness. If you want to build a machine to carry out a particular function (duplicate some human mental activity, for example), you have to understand the connections between the pieces you're putting together and the final operation of the machine. If, as Cohen and Stewart suggest, there is a Gödel-type theorem to the effect that the connection must be so long and complicated that it is impossible for the brain to comprehend, then the builder can't know how to put the parts together to achieve the desired end. This would be an outcome that echoes the results we saw in chapter 13 for chaotic systems, in which it is possible to predict the future in principle but not in practice. It differs from the Divine Calculator scenario in that it doesn't require a new law of nature to block the materialist program, only that the phenomenon of emergence be too complicated to reproduce.

Lest you think that this is a far-fetched possibility, let me tell you about something known as the Enormous Theorem. This is a mathematical theorem involving the classification of mathematical structures called *groups*. Its proof involved more than 100 mathematicians, took over 30 years, and is 15,000 printed pages long. Mathematician Daniel Gorenstein was the man who supervised this work, and when he died in 1992, we may have lost the last human being who understood all of it. Things can, indeed, get complicated in the world of mathematics!

In fact, you can go to a level of speculation far beyond that espoused by Cohen and Stewart. Imagine, if you will, a mathematical system in which there are a series of statements that have to be proved, and the proof of each statement is longer and more complex than that of the last. You could imagine a kind of continuum of such proofs, the limit of which would be a proof that was infinitely long and complicated. The statement corresponding to this proof would be, in essence, unprovable. In the theory of complexity, this would be the analogue of the Gödel Statement in mathematics.

Ways Out

So in the end, there are at least three ways that the theory of complexity might lead us to "impossibility" statements, each of which attacks a different part of the statement, "If we can understand the brain, then we can duplicate it."

It may be that when you get to a system of sufficient complexity, it would become impossible in principle to know what the agents are and what they are doing. This would correspond to a scenario like that outlined above for the Divine Calculator argument. That argument, you will recall, held that if we knew the position and velocity of every particle in the universe, we could use Newton's Laws to predict the entire future, thereby eliminating human free will. It became moot because (among other things) the advent of quantum mechanics showed that it was impossible to obtain the input information. In the same way, the new science of complexity may contain features that make it impossible for us to understand a system as complex as the brain.

On the other hand, when we get to a system of sufficient complexity, we may find that there are laws telling us that we simply cannot build it. This would be a "cyclotron" scenario. Just as the theory of relativity made it impossible to carry through what seemed to be a straightforward building process, the new science of complexity might contain laws that deny the "then" part of the above statement.

Finally, it might turn out that when you put a sufficiently complex system together, you will be unable to predict what its properties are in practice because the connection between the individual parts and the final behavior is too complicated to know. This would be a Gödel-type scenario, as suggested by Cohen and Stewart. I should note that unlike the other two scenarios, this one speaks primarily to the question of whether we could understand a complex machine once it was built, rather than to whether such a machine could be built. There are plenty of examples in the history of technology of people putting together structures without understanding how they worked—the great cathedrals of Europe were built this way, for example.

At our current state of knowledge, there is no way we can say whether any (or all) of these scenarios will happen. If any of them do, however, we will have found our way out of the dilemma posed by the increasing sophistication of the machines we build. We will, furthermore, have done it in a way that preserves both the scientific world view and human uniqueness. In any of these cases, in other words, it would be possible to affirm that the human brain is, at bottom, a material system governed by the same laws as other physical systems, while at the same time being unable to build an artificial brain.

Terminator vs. R. Daneel Ovilaw: What If There Is No Way Out?

Of course, it's always possible that the science of complexity may develop in such a way that none of these scenarios come to pass. It's possible, in other words, that there will be nothing to prevent the completion of the materialist program. What then?

On subjects like this, my experience is that scientists are often the least imaginative thinkers. If you want to get a sense of what the possibilities are, you are much better off turning to fiction and folklore. There is certainly no dearth of tales in which humans create things that act in unexpected ways—think of the Sorcerer's Apprentice, the Golem, or Frankenstein's Monster. In the science fiction of the twentieth century, these sorts of tales tend to center around mechanical creations—robots—equipped with just the kinds of brains that the materialist program would produce. The robots are capable of independent action and are humanlike in virtually all their characteristics (although a common theme is that they cannot or do not have emotions).

The kinds of futures for humanity imagined in the presence of such robots vary from author to author, of course, but there seem to be two general themes: robot-as-threat and robot-as-helpmeet. Let me give you an example of each.

In the classic science fiction movie *Terminator*, the machines turn on their creators and almost succeed in wiping out the human race. The humans fight back and, as the story opens, are on the verge of winning the war. The movie's plot turns on the machines' attempt to send a killer robot (Terminator) back in time to kill the mother of the man who is leading the humans to victory—a classic time travel plot.

In what we can call the Terminator scenario, mankind's ability to create thinking machines is nothing short of a prelude to disaster. The message is clear: Once the machines are built, they will destroy us and the history of the human race will come to a close. And although not all of the Terminator-type stories are as violent as the movie—sometimes the robots simply ignore us and we wither away, for example—the result is always the same. This is the dark view of what a future with thinking machines might be like.

On a more optimistic note, the late Isaac Asimov developed a possible robotic future in which robots became the helpers, and ultimately the salvation of, human beings. In his scenario, when robots were built, the "Three Laws of Robotics" were incorporated into their brains. The Laws* are:

1. A robot may not harm a human being nor, through inaction, allow a human being to come to harm.

2. A robot must obey orders from a human being unless they conflict with the First Law.

3. A robot must protect its own existence unless it conflicts with the First and Second Laws.

In Asimov's novels and stories, the central character is named R. Daneel Ovilaw (the R stands for *Robot*), a robot built to look and behave exactly like a human being. He represents the robot as friend and helper, as a kind of faithful nursemaid for individual humans. In the end, he becomes a Christlike figure whose role is to save and guide the entire human race. He represents the opposite side of the coin from the Terminator—a creature whose enor-

---

*You'd be amazed at how often these three fictional laws crop up in discussions with computer people these days.

mous powers are used to benefit his creators rather than destroy them.

There are, of course, all sorts of fictional futures intermediate between these two. In the *Star Trek* series, for example, there is a robotic character named Data who functions as a pleasantly eccentric colleague in a crew of biological beings, only some of whom are human. You know he's a robot only because of his immense strength and his immense interest in human emotions—an interest that arises because until fairly late in the series, he is incapable of feeling them.

So depending on your mood and general outlook on life, a future involving thinking machines the equal or superior of human beings can be the beginning of the end, the beginning of the millennium, or anything in between. The only statement we can make with certainty is that the real world is still a long way from any of those futures.

The Place of Humanity

But suppose for a moment that one or more of the complexity scenarios we've outlined in this chapter does play out and the materialist program is blocked. What would that imply for the place of humanity in the universe?

We have already seen that a clear separation can be made between human beings and the rest of the animal kingdom, based on our ability to perform specific mental functions. In the last chapter, I pointed out that the new science of complexity allows us to make a case for human uniqueness in terms of the notion of emergent properties. A useful analogy is to think of the evolutionary process as a kind of staircase, with each step corresponding to a new emergent property associated with a new arrangement of neurons. The development of the human cerebral cortex, in this scheme, provides the last step that separates us from our nearest neighbors in the animal kingdom, the chimpanzees.

In just the same way, we suggested that while it is possible to build machines that are intelligent, or even conscious, we have to recognize that these words are being used differently when we apply them to machines and to human beings. A chess-playing machine, for example, just doesn't approach the game the way a human being does. This difference was emphasized in the last chapter by the use of terms such as Intelligence II to refer to a machine like Deep Blue.

I don't think this sort of outcome would be troublesome to most people. The ability to build tools, after all, has always been one of humanity's distinguishing features. We build vehicles, but we don't feel diminished because our cars can "run" faster than we can. No one, for example, is suggesting that we should cancel the Olympics because we have the Indianapolis 500. A machine that can play chess but doesn't possess consciousness would, it seems to me, be in the same nonthreatening category.

If we think about machine intelligence in this way, then our attention naturally turns to delineating the differences between the Intelligences and Consciousnesses associated with the different Roman numerals. It seems likely to me that the things we intuitively associate with human uniqueness—emotions, for example, or the ability to develop moral systems—may well turn out to be precisely those qualities that differentiate Consciousness I from all of the rest. Whether this is true or not, however, moves the question of how we are different from machines from philosophy to observational science, and this can only raise the relevance of the debate.

There is a visual metaphor we can use to talk about the role of humanity in a world of dissimilar mentalities. In this metaphor, humanity is still standing at the top of an evolutionary stairway, with each step representing a different emergent property of the brain. We can include machines in this analogy by thinking of something like Deep Blue as being on top of a distant mesa, a mesa we might label Intelligence II. In fact, it's not too hard to imagine that we will eventually build many such machines, each inhabiting its own mesa in our landscape, each with its own Roman numeral.

From the top of our stairway, we will look out across the landscape and see ourselves as unique products of evolution, similar to, but different from, every other kind of intelligence and consciousness in the universe. We will also understand that, while the stairway we stand on was built by nature, we are, ourselves, in charge of building the "Monument Valley" of mesas that surrounds us.

There will, after all, be something left for us.

Bibliography

One of the most engaging characteristics of the debate over the nature of consciousness is that a lot of it is being carried out in books that are written for, and accessible to, the general public. I can't think of any time since the period following Darwin's publication of *The Origin of Species* that this has been true. Here are some books that I enjoyed and that you might find useful.

Cohen, Jack, and Ian Stewart. *The Collapse of Chaos: Discovering Simplicity in a Complex World*. New York: Viking Penguin, 1994. If the world is really complex, nonlinear, and chaotic, why does it seem so simple? The authors (a developmental biologist and mathematician, respectively) tackle this problem from a variety of viewpoints. My favorite was the discussion of the DNA code that pointed out that the code is useless without a machine to read it (a "machine" we usually call a mother). The book gets a little cutesy at times, but it's worth ignoring that problem to get the wide range and deep discussion of everything else.

Crick, Francis. *The Astonishing Hypothesis*. New York: Simon and Schuster, 1994. Crick, who shared the Nobel Prize for discovery of the structure of DNA, has never been known for being shy about stating his views. This book is a straightforward statement of the materialist position on brain function, backed by a lot of information on current brain research. Some of the explanations may get a bit detailed for the general reader, but stay with them. They will repay your effort.

Damasio, Antonio R. *Descartes' Error: Emotion, Reason, and the Laws of Physics*. New York: Avon Books, 1994. Neurophysiologist Damasio talks about the brain as a biological (as opposed to computational) system, with a lot of discussion of case histories and laboratory experiments.

His main argument is that you can't really separate the brain from the rest of the body, as Descartes did, and I think he makes the point quite well.

Dennett, Daniel C. *Consciousness Explained*. Boston: Little, Brown, 1991. This is one of the seminal books in consciousness studies. It's easy to read, and even if you don't agree with some of Dennett's arguments (as I don't), you have to read it if you want to have a sense of the current debate.

Kauffman, Stuart. *At Home in the Universe: The Search for the Laws of Self-Organization and Complexity*. New York: Oxford University Press, 1995. The thesis of this book is that complex systems will display order spontaneously (we used the term "emergent phenomenon" to talk about this in the text). It gives as good a description of the phenomenon as you're likely to find anywhere. My problem is that the next step—the one connecting the complex systems generated in computers to living things—seems to be missing.

Nagle, Ernest, and James R. Newman. *Gödel's Proof*. New York: University Press, 1958. This book is one of those classics—a little-known jewel that covers a small topic as well as it can be covered, and that therefore enjoys a kind of underground fame among cognoscenti. If you want to know more about Gödel's Theorem, this is the place to go.

Penrose, Roger. *The Emperor's New Mind*. New York: Oxford University Press, 1989.

———. *Shadows of the Mind*. New York: Oxford University Press, 1994. As outlined in the text, Penrose uses these books to advance two arguments: first, that because of Gödel's Theorem, it is possible to prove that the brain is not a computer, and, second, that the essence of understanding the brain lies in the new science at the nexus of quantum mechanics and unified field theories. This is deep stuff, and you should be warned that the books are often *very* heavy going. Even with a Ph.D. in theoretical physics, I had to make a real effort to understand some parts. Fair warning.

Pinker, Steven. *The Language Instinct*. New York: William Morrow, 1994. One of the nation's leading young linguists tells the story of language and argues for the existence of "grammar genes." There's a lot of meat here, including exercises in sentence diagramming and word usage that may be a bit daunting for some. I have to admit, though, that my original favorable reaction to the book, generated by Pinker's use of soap opera dialogue to illustrate the role of language, was amply borne out by his subsequent writing.

Restak, Richard. *Brainscapes*. New York: Hyperion, 1995. Restak is without doubt the best author for the general reader to consult on the subject of the brain and its functions. His books *The Brain, The Mind*, and *Receptors* (Bantam Books, 1984, 1988, and 1994, respectively) give a wonderfully complete discussion of the subject. The last book in particular covers the new chemical understanding of the brain. *Brainscapes* is a short (135 pages) summary of our current understanding, and I would recommend it for those wanting a quick overview. Maybe after you've read it, you'll want to go on to the more detailed discussions.

Savage-Rumbaugh, Sue, and Roger Lewin. *Kanzi: The Ape at the Brink of the Human Mind*. New York: John Wiley & Sons, 1994. A firsthand account of language ability in chimpanzees and the training of the bonobo chimpanzee named Kanzi (see chapter 4). Even if the subject doesn't interest you, the story of Savage-Rumbaugh discovering her primate roots on a deserted beach in Portugal makes this book worth the price of admission.

Veggeberg, Scott. *Medications of the Mind*. New York: Henry Holt and Co., 1996. A short, concise, and very readable treatment of the new field of molecular psychiatry, including details on the way that drugs like Prozac affect the neurons in the brain. This is as good an overview of this field as you're going to find.

Acknowledgments

Many people have been kind enough to comment on the manuscript for this book. With the usual caveat that whatever mistakes remain are my responsibility alone, I would like to thank David Aitel, Ann Butler, Kenneth De Jong, William Finnerty, Larry Hunter, Kathryn Laskey, Ryszard Michalski, Harold Morowitz, Jeffrey Newmeyer, and James Pfiffner for their help. I would particularly like to thank Flora Waples-Trefil for assistance in preparing the manuscript and for the kind of criticism that only a teenage daughter can supply.

Index

About the Author

Coauthor of the bestselling book *The Dictionary of Cultural Literacy* and the highly acclaimed *Science Matters*, and author of *1001 Things Everyone Should Know About Science*, James Trefil has published more than twenty books on science. A former Guggenheim fellow and a regular guest on National Public Radio, he is the Robinson Professor of Physics at George Mason University and a contributing editor of *Smithsonian* magazine. He lives in Annandale, Virginia.